14488

Imperial Airways

Imperial Airways
and
the first British airlines
1919–40

A. S. JACKSON

TERENCE DALTON LIMITED

Published by
TERENCE DALTON LIMITED

ISBN 0 86138 098 3

Printed in Great Britain at
The Lavenham Press Limited, Lavenham, Suffolk

Contents

Acknowledgements		vi
Foreword		vii
Chapter One	The Early Birds	1
Chapter Two	The Chosen Instrument	11
Chapter Three	The European Services 1924–28	25
Chapter Four	Flights to India 1921–32	39
Chapter Five	Cairo to the Cape, 1919–34	52
Chapter Six	The Growth of Domestic Competition	61
Chapter Seven	Australia and the Far East	79
Chapter Eight	The South Atlantic, Portugal and West Africa	89
Chapter Nine	The Empire Flying-boats	95
Chapter Ten	The North Atlantic Challenge	102
Chapter Eleven	Mr. Perkins demands an Inquiry	117
Chapter Twelve	The Cadman Report	129
Chapter Thirteen	The Short Reign of Sir John Reith	139
Epilogue		146

Appendices

Appendix One	Airliners operated by Imperial Airways 1919–40	147
Appendix Two	Details of airliners operated by Imperial Airways	152
Appendix Three	Management of Imperial Airways and British Airways	154
Appendix Four	Secretaries of State for Air 1919–40	155
Appendix Five	Captain Hinchliffe and the Hon. Elsie Mackay	156
Bibliography		157
Index		158

Acknowledgements

I am most grateful to those who helped me with photographs for this book. They include Lord Brabazon of Tara, who also kindly wrote the foreword, Mrs. L. Bennett, Mr. M. Cobham, Capt. V. Hodgkinson, Mr. R. Jackson, Mr. D. Munro, Mr. C. Reith, Mr. T. Samson, Mr. J. Stroud, British Airways, the Boeing Aircraft Company, The FAA Museum, the RAF Museum and Lufthansa.

Essential information was provided by Mrs. J. Blockley, Mr. A. Finch, Miss J. Hinchliffe, Air Commodore Powell, Captain J. Scott and the British Airways archivists, Messrs. K. Hayward, R. Huntley and R. A. R. Wilson.

I also wish to thank many others whose interest has contributed to this book, not least Mrs. Pamela Ansell, who typed the manuscript.

For Tracey

Foreword

by Lord Brabazon of Tara

To those of us interested in the development of British civil aviation, from its beginnings until the outbreak of the second world war more or less stopped further growth, the story of British airlines (and especially Imperial Airways) is a particularly fascinating one.

We are therefore indebted to Captain Jackson for this opportune and absorbing account of those pioneering years, starting with the early cross-Channel flights and tracing the development of services to nearly all parts of the empire and, almost, across the North Atlantic.

What is surprising is how many of the problems faced today in civil aviation were the same faced in those early years: the difficult negotiations for traffic rights, combating the advantages of subsidised foreign airlines, the thorny questions of the extent of government involvement and regulation. However, this book reminds us vividly of the bravery and skills of the pilots and crews (and indeed their passengers) as they guided their fragile machines over sometimes uncharted territory. Today we fly non-stop between London and Singapore or Los Angeles and think nothing of it – this book is an opportune reminder of how much we owe those brave pioneers.

Overleaf: Handley Page's converted bomber 0/400 at Cricklewood aerodrome. 1919.
Croydon Airport Society

CHAPTER ONE

The Early Birds

Aviation was in its infancy when the First World War broke out in 1914. Five years earlier Blériot had flown across the English Channel. The offer by Lord Northcliffe of £10,000 to the first pilot to fly across the North Atlantic Ocean seemed unlikely to be taken up for very many years. Yet inevitably the need to provide the fighting services with the best possible equipment led to an improvement in the quality of both aircraft and their engines.

The chief designer of the Aircraft Manufacturing Company was Geoffrey de Havilland. In 1916 a subsidiary was acquired, Aircraft Transport and Travel. The founder of AT & T was George Holt Thomas who had built Farman aeroplanes under licence from the French parent company. He anticipated a bright future for civil aviation after the war. His dream was to establish an airline "to enter into contract for the carriage of mails, passengers, goods and cattle." He had not foreseen the obstacles that would be placed in front of airlines wishing to carry passengers across the territories of other nations.

Lord Northcliffe was the proprietor of *The Times* and the *Daily Mail* and he had used these newspapers to campaign for the replacement of the Prime Minister, Herbert Asquith, by his fellow Liberal, David Lloyd George, in 1916. Thereafter the war was prosecuted with much more vigour. In May 1917 Northcliffe was made chairman of a committee to study "the development and regulation after the war, of aviation for civil and commercial purposes from a domestic, an imperial and an international stand-point".

Northcliffe had never flown but several of the members of his committee had done so. Sir Sefton Brancker was in charge of personnel for the Royal Air Force. Captain J. C. Porte was the Royal Navy's flying-boat specialist. Brigadier Maitland commanded the British Military Airship Service and was to travel aboard the R34 on its journey across the North Atlantic and back in 1919, the year in which Northcliffe's prize was won by Alcock and Whitten-Brown in a Vickers Vimy machine. Among the forty members of the main committee was the writer H. G. Wells.

It was agreed that an International Convention for Aerial Navigation should be convened and the attitude of the British government to freedom of the skies defined. In 1910, at the first ICAN conference in Paris, the Foreign Office had followed traditional maritime practice and declared the air above the nation's territories, both at home and abroad, to be inviolable.

In 1919 Germany was specifically excluded from the deliberations of the Peace Conference, but was the only country pressing for freedom of the air. In the event the committee's recommendation was that Great Britain and the Dominions should stand by the decision taken before the war.

A member who dissented most strongly from this view was Frank Pick. He pointed out that Britain's position as a maritime power bore no relation to her commercial and strategic situation. Not only could Germany deny over-flights of British aircraft, but France, Spain and Italy could prevent the extension of air routes across Europe to the Empire. In order to get freedom of passage for her aircraft it was imperative for Britain to have such international questions settled. Pick also disagreed with the decision to put civil aviation under the control of the Air Ministry, a body instituted in 1917. He reminded the committee that the Merchant Marine was supervised by the Board of Trade and not by the Admiralty, although both had responsibilities for ships.

H.G. Wells also made a useful contribution in a minority report in which he argued:

> The British Islands are small islands and our people numerically a little people. Their only claim to world importance depends upon their courage and enterprise, and a people who will not stand up to the necessity of air service planned on a world scale, and taking over thousands of aeroplanes and thousands of men from the onset of peace, has no business to pretend to anything more than a second rate position in the world. We cannot be both Imperial and mean.

He went on to deplore

> the narrowness of outlook that has debarred the committee from seizing its opportunity to plan an air service not only great in scale but great in spirit.

With greater sense the committee suggested that air transport should be developed to create a market for the country's aircraft industry, which would in turn produce more efficient machines. In its opinion government assistance would be required to sustain the industry during the initial stage, civil aviation developing as a private enterprise. The viewpoint of the RAF was expressed by Lord Trenchard whose keenness for a thriving civil aviation sector was based upon his wish to have a reserve for his own service. He too advocated the allocation of public money for the development of routes to the Empire, on the grounds both of strategy and prestige.

The government took very little notice of this advice and adamantly refused to make a subsidy. It did however provide aerodromes with hangar space, emergency landing grounds, meteorological information, and a free wireless communication service. Landings and aircraft storage involved fees to be paid by the aircraft owner. Early in 1919 the Department of Civil Aviation was created at the Air Ministry. Major-General Sykes was appointed

Controller-General with direct access to the Secretary of State for Air. This post was held by Winston Churchill but without cabinet rank.

In May 1919 the Air Ministry began to supervise the issue of 'B' licences to commercial pilots and Certificates of Airworthiness for civil aircraft. In July it was announced that Hounslow aerodrome would be the required stop for Customs clearance for flights entering and leaving England. Hounslow did not possess a wireless station and pilots learnt to find it by first identifying the bridges crossing the Thames river, which lay adjacent to the grass airfield. The only indication of its international significance was the word CUSTOMS on one shed and DOUANE on the other. After sunset a gas-burning rotating beacon shone dimly for a few hours. A shed served as a waiting room but no refreshments could be obtained.

To coincide with the peace celebrations in Paris the Air Ministry advised that, for one week from 14th July 1919, flying to and from France would be authorised. Colonel Pilkington, a member of the St. Helens glass manufac-turing family, noticed the announcement in the *Evening News* and immedi-ately telephoned AT & T to charter an aircraft to fly him to the French capital. After a few minutes discussion a one-way fare of £42 was agreed. Although Sir Sefton Brancker had been appointed Managing Director no advertisements of regular services had yet been published. The military DH4 and DH9 machines which they possessed had been converted for civil use and were designated Airco 4 and Airco 9.

AT & T's pilot, Jerry Shaw, supervised the kitting out of his passenger in a heavy leather coat, helmet, goggles and gloves, and flew out of Hendon at 7.30 a.m. on a day of torrential rain and low cloud. Anxious to be the first to reach Paris with a paying passenger Shaw flew over Hounslow, but seeing another machine taking off, he decided not to land there and continued his flight to le Bourget. Describing this trip he wrote:

> Apart from a few tricky minutes in low cloud near the North Downs the journey over Folkestone and Boulogne down to Beauvais was uneventful but wet and hardly ever over 200 feet above ground ... we eventually landed at le Bourget at 10.15 a.m. In those days the airfield consisted of several canvas hangars, some wooden sheds and a lot of mud.

Le Bourget was also the official Customs port of entry, but Shaw saw no officials about and no one showed any interest in his arrival. Not possessing a passport he and Colonel Pilkington obtained the assistance of a French Air Force officer to leave the airport premises and they boarded a tram for Paris. The following day the two men flew back to England, landing at Hounslow, where an irate Customs officer listed the number of regulations which Shaw had broken.

In August AT & T began scheduled services to Paris and by 1st November, after ten weeks of operation, 147 flights had been completed.

Leather helmets, coats and gloves were needed by passengers flying in a DH.9.b in 1919. *Croydon Airport Society.*

The single fare was £21 and passengers were also offered road transport from central London to Hounslow and from le Bourget to Paris. From the start the company faced competition.

Handley Page Transport had been founded in 1908 and during the war had built the largest bombers, the O/100 and O/400. In 1919 their workshops at Cricklewood were almost deserted. The cancellation of war contracts had been followed by dismissal of the workforce. With one research engineer and a test pilot Sir Frederick Handley Page organised the conversion of his machines to civil use. His advertisement for a General Manager was answered by Major George Woods Humphery, who had been a pilot in the Royal Flying Corps. Major H. G. Brackley was appointed Air Superintendent. In an unsuccessful attempt to win Lord Northcliffe's £10,000 prize he had been provided with a Handley Page bomber. The chief pilot's post went to Lieutenant-Colonel Sholto Douglas, who many years later became chairman of British European Airways. Another of the pilots to be employed was Gordon Olley, who subsequently owned his own airline.

The first commercial flight of Handley Page Transport was made by Sholto Douglas from the works airfield at Cricklewood. On 22nd July 1919 he flew a chartered Handley Page O/400 to Brussels. Regular services to Paris began on 31st August. In September a scheduled service to Brussels was

A former Royal Flying Corps fighter pilot Gordon Olley flew with Handley Page Transport before joining Imperial Airways. In 1934 he founded Olley Air Service. *R. A. R. Wilson*

inaugurated. The O/400 was a biplane powered by two Rolls-Royce Eagle engines. The company's publicity always stressed that their aircraft had two engines. In fact the pilot had no choice but to land as soon as possible if one of these failed, as they could not maintain altitude with only one engine in action.

Handley Page offered the ten passengers who could be carried in their aircraft a limousine service to and from the aerodromes. The fare to Paris was the same as the first class rail and boat fare and the flight took about three hours. A box lunch of sandwiches, fruit and chocolate was available on payment of three shillings. When the larger W4 was introduced on the route twenty passengers could be carried in a little over two hours. Their aircraft also boasted a lavatory and in one version an enclosed cockpit for the pilot. After a very bumpy ride to Paris the pilot, W. Hope, ordered a mechanic to dismantle the canopy, explaining that he liked to feel the wind on his face. When he returned to see the result he was exasperated to observe that very little progress had been made. He went away and reappeared with an axe and swiftly demolished what remained of the canopy.

In the autumn of 1919 a railway strike encouraged the airline to increase the frequency of its service. Revenue also increased from the carriage of 250,000 letters between the two countries. That year a third challenger appeared. Sir Samuel Instone and his brother were shipowners, shipbrokers and coal exporters, with their headquarters in Wales. Their operations were being hampered by postal delays that led to their steamers lying up for as many as ten days in French ports, unable to discharge because the bills of lading had not arrived. Their solution was to obtain an aeroplane and a pilot, thereby saving a huge sum in ship hire and demurrage. From this small beginning The Instone Air Line entered the competition in October 1919 when their pilot, Franklyn Barnard, flew a converted Vimy bomber from Hounslow to le Bourget. Passing over Boulogne he threw out a number of greetings cards, decorated with the Union Jack and Tricolour, and bearing the address of Instone's office in Paris. Before long the company began a service to Brussels, and in April 1920 a flight was made to Prague to investigate the possibilities of that route.

By the end of 1919 the three British companies also faced competition from two French airlines and Holland's KLM. With at least four cross-Channel services every day, Hounslow's grass airfield was looked upon as a thriving airport. When Customs consented to install an office at Crickle-wood, Handley Page were able to operate to the continent without first flying their aircraft empty to Hounslow. But the grass field was short and pilots thought themselves lucky to be airborne with more than a hundred yards remaining, as they aimed for the space between two hangars on the airfield perimeter. Moments later their aircraft would pass about fifty feet above

Cricklewood Broadway. An engine failure at that stage would have been disastrous. Handley Page's pilot, R. H. McIntosh observed that the initial route to the Continent took him along the Edgware Road, plied by Number 10 buses. He found he could only overtake them when they stopped for passengers. He was able to breathe more freely when his aircraft passed over Marble Arch at a height of 500 feet.

Meanwhile Handley Page's new type W8 airliner was ready for flight testing. The test pilot was ill and Brackley in the USA, so Sholto Douglas was ordered to take the tests. Under the circumstances he asked for extra pay. This was refused and he resigned in disgust and rejoined the RAF.

The weather was as much a hazard as the reliability of the machines. Blind-flying instruments had not been developed, altimeters could not be relied upon and compass needles oscillated wildly except when a steady course was being flown in calm conditions. Late in October it was snowing when Jerry Shaw left le Bourget with one passenger bound for London. His compass failed and with fuel running low he was obliged to ditch alongside a small vessel in the Channel. The passenger, bowler hatted, strode unconcernedly along the wing to be pulled on to the coaster. He and Shaw were put ashore at Weymouth the following morning, and completed their journey to London on a train.

The first Director of Civil Aviation, Major-General Sir Frederick Sykes, wrote at the end of 1919:

> It may be questioned whether civil aviation in England is to be regarded as one of those industries which is unable to stand on its own feet, and is yet so essential to the national welfare that it must be kept alive at all costs.

He recommended both government subsidies and help in kind so that the British airlines would be able to survive against the heavily subsidised foreign competition. Once again his words fell on deaf ears.

In Belgium, Denmark and Sweden the airlines were state monopolies. This was also the case in Holland, a country which had not been ravaged by war. Germany, denied an air force by the terms of the Treaty of Versailles, backed several companies, and the larger towns paid these to use their airports. The Germans also managed a company which operated on behalf of Russia between Königsberg and Moscow. Subsequently this route was extended to Teheran and Peking. France was fearful that a revived Germany would once again pose a threat to the peace of Europe, particularly as the United States government had refused to ratify the former President Wilson's pledge of military assistance in such an event. The British government had also withdrawn a similar pledge. Consequently the French retained large military forces and supported their airline companies as a reserve for their air force.

IMPERIAL AIRWAYS

In March 1920 another area of grass fields replaced Hounslow as the Customs airport for London. This combined the former RAF airfield at Wallington with another called Waddon. The flying area was on Waddon, rough, narrow and so uneven that aircraft could be lost to sight from the tarmac apron on the Wallington side. An old factory and a water tower on the perimeter presented dangerous obstructions and pilots encountered a sudden downdraught when a south-west wind was blowing over the Purley valley. Sheep sometimes had to be shepherded away from the landing area. Initially the airport had no control tower until one was built on stilts over the Customs shed. Lacking radio the controller used an Aldis lamp to show a green or red signal to pilots wishing to land. When an enterprising hotel chain applied for a licence to convert the canteen into a hotel the presiding magistrate asked whether alcohol was required to revive terrified passengers. Lacking office accommodation a haphazard building programme generated rows of wooden huts of all shapes and sizes. The nearest railway station was at Waddon so passengers depended upon private cars or airline limousines. Those airport workers who did not have bicycles had to walk a considerable distance from the station. The residents of nearby Wallington had no wish to be associated with London's new airport so it was adopted by Croydon.

The entrance to Croydon Aerodrome in 1920.

That month a third French company began services between Paris and London. Subsidies also enabled the French to start a service to Morocco with a stop in Spain, carrying passengers and mail. Twice a week they operated to Turkey. An inaugural flight to Dakar in French West Africa had been made with refuelling stops in Algiers and Timbuktu. Struggling against the competition in Europe the British companies could not contemplate extending their services to the Empire.

Although the Peace Conference had been followed by an Air Navigation Act in 1920, guaranteeing freedom of passage for aircraft, the rivalry among the European countries prevented this happening. Germany blocked the route to Prague, while Italy and France prevented the British from flying over their territory until the late 1930s.

Another government committee, chaired by Lord Weir, recommended financial support for the companies on the cross-Channel route. Sir Sefton Brancker pointed out that increased subsidies by the French government had enabled their airlines to reduce the London–Paris fare to six guineas (£6.6s). Warsaw was about to be added to their network. One German company had already flown more services than all the British airlines put together. Yet the British government remained deaf to all the arguments: "Civil aviation must fly by itself," Winston Churchill told the House of Commons. "The Government cannot possibly hold it up in the air."

Nor was the GPO willing to charter and pay for a fixed amount of space, something which it had done for the shipping companies. In 1839, to encourage a policy of larger, faster steamships Mr. Cunard had been paid a subsidy. The GPO would only pay the airlines for space as and when required. Even at fivepence an ounce the public showed very little interest in sending their letters by air. By contrast, the air mail service started in the USA in 1918 quickly became economic and highly reliable, so that within months charges were virtually halved.

Sadly civil aviation and the prospects for its future had not caught the imagination of the public, which was in no mood to pay taxes for the benefit of well-to-do travellers. The political parties shared a common desire to reduce the high levels of taxation that had remained in force since the end of the war. There was little recognition that air power might one day supersede naval supremacy as a powerful weapon in the defence of the British Isles. A return to 'business as usual', free trade and the economies to be obtained by disarmament, were the priorities of the day. Yet 1920 was a boom year as the long established manufacturers of textiles, iron and steel, and ships flourished. Exports were double those of 1913.

Although the British pilots had been flying single-engined aircraft, most of them converted air force machines, frequently without adequate weather reports and in every kind of weather, only two fatal accidents in scheduled operations had occurred by the end of 1920. But financially the three companies were reaching the end of their resources. George Holt Thomas put AT & T into liquidation. The Instone Air Line could only operate to Paris when there was a sufficient demand for seats. Only Handley Page Transport kept flying a regular service on that route.

On 28th February 1921 all three companies were forced to cease operations, leaving the European airlines to fly into Croydon on their own. Thus the British taxpayers were contributing to the upkeep of aerodromes, wireless and weather services, used only by foreign concerns. Major Brackley left Handley Page Transport to join an official mission to instruct the Japanese naval air arm. His replacement as chief pilot was McIntosh, who along with W. Rogers, offered Sir Frederick their services without pay until a new system of funding allowed flights to be resumed. This generous suggestion was declined with regret, their employer attributing the shutdown to the government's lack of interest in civil aviation.

In a letter to *The Times* Sir Sefton Brancker complained that in 1919 Britain's airlines had possessed the pilots and aircraft to offer services throughout northern Europe. The government had failed to provide a guarantee of assistance at the Hague, where AT & T had been negotiating for the organisation of a through service from London to Stockholm and Oslo with halts at the Hague, Hamburg and Copenhagen. As Holland, Denmark, Norway and Sweden had few pilots and no aircraft of their own manufacture, most of their airlines could have been manned and equipped from England. "The British firm," he wrote, "has now gone into liquidation. The result is that Mr. Fokker, the designer of the most efficient fighting machine which Germany produced during the war, is to have the order for the equipment."

Meanwhile in the absence of any British air service it was left to the French to handle mail sent by air from the United Kingdom. An editorial in *The Aeroplane* summed it up: "British civil aviation died with the cessation of the Handley Page cross-Channel service, killed by the forward policy of the French government and the apathy of our own."

Another blow to British hopes for progress in aviation was the loss of the new airship R38 in 1921, together with the lives of Air Commodore Maitland and the whole crew. Subsequently the government decided to order two airships for long distance travel. The R100 was built by private interests, the R101 at the Royal Airship Works at Cardington in Bedfordshire. However, it was 1929 before either was ready for an international proving flight.

CHAPTER TWO

The Chosen Instrument

The British government obviously had to do something and on 9th March 1921 Winston Churchill announced the appointment of the Air Ministry Cross-Channel Committee to decide how to revive air services. There was a coal miners' strike before the committee reported and, when it seemed likely that the railwaymen and transport workers would also come out in sympathy, a national emergency was declared. At this juncture Captain Frederick Guest was appointed full-time Secretary of State for Air.

A 'temporary' subsidy was speedily arranged to take effect on 19th March and the Air Ministry asked for tenders. Woods Humphery, temporarily unemployed, tried to get six individuals to put up £5,000 each to form a new company. He was joined in this endeavour by Colonel Frank Searle, organiser of the London General Omnibus Company. They were not successful but Searle was able to persuade the BSA-Daimler Group to put up £30,000. The Air Ministry accepted the tender of a newly constituted Daimler Airway, and loaned some DH18s which had been intended for AT & T. Searle was appointed Managing Director with Woods Humphery as General Manager. The two men adapted the Daimler Hire system of intensive utilisation of automobiles, and aimed to obtain 1,000 hours flying time by each aircraft annually. Searle had found that no aircraft made more than one round trip to Paris in a day, thereafter being put in a hangar for checks which could be completed quite quickly. In May a Daimler aircraft made two return journeys to Paris in one day, the first time a commercial airliner had done so.

Meanwhile subsidies had been granted to Handley Page and Instone, the object being to guarantee them a profit of ten per cent on gross receipts. Each was to receive £75 per flight with a maximum liability to the Air Ministry of £88,200 in the financial year ending 31st March 1922. In fact the companies received rather less than this sum, while over the same period the French government subsidised their own companies to the tune of £1,328,600.

Even before the airlines resumed services some of their aircraft had been fitted with Marconi wireless sets. Range of reception was about fifty miles. Handley Page were the first to use the equipment, soon to be followed by Instone. In 1921 the pilot, McIntosh was *en route* to Croydon which was shrouded in fog. One of his passengers was the editor of the *Daily Chronicle*.

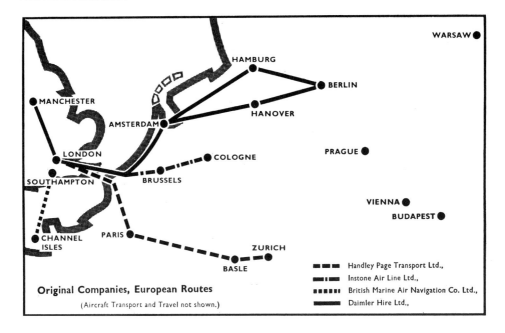

Original Companies, European Routes
(Aircraft Transport and Travel not shown.)

McIntosh was unable to use his set because the trailing aerial had caught up in the branches of a tree and broken, but he had been able to see the two towers of the Crystal Palace and, knowing the course to be flown from there, was able to land safely at Croydon. Reporters were kept in ignorance of this and the front page headlines in the following day's newspapers dubbed the pilot 'All-weather Mac' and welcomed the advantages of wireless telegraphy.

When the weather at Croydon was too bad for pilots to attempt a landing the controller at Lympne was informed by telephone and he fired a Very light to warn pilots flying from the Continent that they should proceed no further, but land on his airfield. The passengers were put on a train for London. Other emergency airfields were Marden near Ashford and Penshurst near Tonbridge. Each of these had a resident caretaker, a telephone and supplies of petrol and oil.

To assist pilots to navigate, some railway stations in southern England had their names painted in white on their roofs. At Edenbridge the name was cut in the chalk near by. It was normal practice to navigate by reference to railway lines. Beacons had been installed at six points in England and nine in France to guide pilots along the right route, each beacon flashing a distinctive sign in Morse code.

The limitations of these measures were revealed in April 1922 when a DH18 of Daimler Airway collided head-on with a Farman Goliath of a

French company. There was fog at the time and it was believed that both pilots were looking down upon the Abbeville–Beauvais road for lack of forward visibility. Even on a clear day the pilot of a DH18 suffered the disadvantage of a cockpit situated aft of the mainplanes. In the House of Commons questions were asked about the efficiency of the weather and wireless service. Dual controls on aircraft were suggested. The newspapers expressed great dismay at the lack of adequate assistance to pilots.

That same month an inappropriately named 'permanent' subsidy scheme was introduced. The Air Ministry authorised the three companies, which now included Daimler Airway, to operate in competition to Paris, with a subsidy of 25 per cent of gross earnings plus £3 per passenger carried and threepence per pound weight of goods and mail. The Air Council provided half of the aircraft on a hire-purchase basis and paid half of the premiums on the insurance taken out by the companies. The Instone Air Line was further subsidised for the London–Brussels service by a sum not to exceed £20,000 a year.

These arrangements were made at a time when the boom of 1920 had given way to a severe depression. Orders for ships to replace those sunk during the war had fallen off drastically; the demand for iron and steel diminished; coal was suffering from competition from oil and electricity generation, and the textile industry saw its overseas markets suffering severely from the new suppliers led by Japan. The British government appointed Sir Eric Geddes, the Chairman of the Dunlop Rubber Company to wield an 'axe' on its current and future public expenditure programme. The 1922/23 Air Estimates included a cut in the Civil Aviation vote. The Department of the Controller-General of Civil Aviation was down-graded to a Directorate. The incumbent resigned and was replaced by Sir Sefton Brancker, who had been Head of the Air League of the British Empire since the demise of AT & T.

In this crisis Brancker urged the government to adopt an Empire air mail scheme, and recommended the establishment of one national airline to compete against the Europeans. Theodore Instone thought it absurd to subsidise three British airlines on the same route, London to Paris, where there was already strong foreign competition. In May 1922 there were seven British flights a day to Paris, each of which qualified for a subsidy. One day during June a DH34 made five single journeys between Croydon and Paris. Captain Guest, Secretary of State for Air, although without a seat in the cabinet, believed that public money should only be allocated to an airline whose European services provided links with the Empire.

In the summer of 1922 traffic was well below expectations and the three airlines complained that they could not earn the full subsidy. In October another 'revised' subsidy scheme was introduced. The government agreed to drop its insistence on the British airlines competing against each other, and routes were allocated to each one in the hope that they would develop new extensions. Equipment on hire-purchase was made a free gift. Subsidies for the period from 1st October 1922 to the end of March 1924 were granted on a lump sum basis.

Daimler Airway received £82,500 and agreed to inaugurate a new service from Manchester to London and Amsterdam. A further extension to Berlin was planned, but was blocked by the German for many months. Handley Page received £22,500 and kept the London to Paris route with an extension to Lausanne. The Instone Air Line retained London to Brussels with a subsidy of £37,500. Their wish to extend to Prague with a stop at Cologne was obstructed by the Germans. A subsidy of £10,000 was also offered to a newcomer, The British Marine Air Navigation Company, which proposed to operate amphibious Supermarine Sea Eagles from South-ampton to the Channel Islands and to one or more French ports. However

Sir Sefton Brancker. Cartoon by Charles C. Dickson in the Pilots' room of the Aerodome Hotel.

operations did not begin until 1923. Difficult landing conditions made the service seasonal and the French were disinclined to permit landings on their waters.

These government schemes never appeared likely to permit civil aviation to 'fly by itself', but did have a beneficial effect. By the end of 1922 just over 65 per cent of the passengers and 41 per cent of the cargo was being carried by the British on the route to Paris. The airlines tried to inspire some national interest through publicity by arranging press flights, tours of Croydon Airport and joy rides costing five shillings. The Instone Air Line gave names to its aircraft and decorated them in the company's livery of silver wings and a royal blue fuselage, their name painted on each side. This was the first airline to provide their pilots with a nautical blue uniform which set a pattern adopted by very many other airlines in the world. The three carriers benefited from the stability deriving from the assured government money and the freedom to carry only the most remunerative traffic. But, compared to the growing European airlines, the total number of staff employed was derisory: eighteen pilots and 117 other employees.

Daimler Airway employed four pilots and operated only four aircraft. This was the first airline to employ a steward. He wore a stiff wing collar with bow tie and monkey jacket. The DH34 in which he flew was not heated so the passengers often sat wearing their overcoats. Early in 1923 Woods Humphery flew to Berlin to arrange a Daimler service to the city. This was followed by the arrival of the first German aircraft to land in England, bringing a team of negotiators. Seen as a welcome sign of a more co-operative spirit it may have been brought about by the merger of the two largest French companies, which then reduced the London to Paris fare to £4.10s

Initially Instone envisaged a weekly flight connecting at Amsterdam with the German company Aero Lloyd. Eventually a North European Grand Trunk Airway was formed. This also involved KLM and a Danish company and was so popular that the exchange of passengers at Amsterdam was abandoned in favour of through flights. To ensure a seat it was advisable to book two weeks in advance. Instone also benefited from the French occupation of the Ruhr in 1923. This had been France's response to Germany's default on the payment of reparations. The only way the German government could pay the huge number of unemployed workmen was to charter the British airline to fly currency to London and then back to Cologne. This earned Instone £120 a trip.

Meanwhile in Britain there had been a change of government. The Conservative Party had ceased to regard Lloyd George as worthy of support as leader of the Coalition. On the contrary he was seen to have become a liability, and a Conservative government under Bonar Law had taken office late in 1922. The new Secretary of State for Air, Sir Samuel Hoare, appointed Sir Herbert Hambling and two other business friends to look into the existing method of subsidy, and to propose a scheme which would give the greatest permanent benefit to the airlines at the least cost to the taxpayers. Bonar Law's advice to Sir Herbert was to persuade the companies to amalgamate.

While the Hambling Committee was doing its work Stanley Baldwin replaced the ailing Bonar Law as Prime Minister. Sir Samuel Hoare became the first Secretary of State for Air to be accorded a place in the cabinet. He did not have to wait long for the committee's report. In its opinion the foreign companies with their generous subsidies were providing more than enough competition for the British airlines. Therefore it was wasteful for the latter to compete among themselves. Nor could they operate economically on the short routes to Paris and Brussels. Longer routes would have to be developed. The committee clearly favoured the creation of one monopoly company with the interests of the existing airlines safeguarded. This company should be run as a commercial organisation on business lines and should not be under government control. A private concern with large resources of its own would have to accept the risks of expenditure on equipment and the development of routes. Additionally the government should be bold enough to accept the need for a total subsidy of a stated amount to be spread over a period of ten years. Consequently someone at the Air Ministry should be responsible for relations between the government and the monopoly company.

Nine months of bargaining passed before the financial arrangements were settled. Sir Samuel Hoare later wrote:

> The Treasury did not believe in civil aviation and strongly objected to long term commitments to companies that were obviously in financial difficulties . . . The solution was for Szarvasy's British, Foreign and Colonial Corporation to put up the capital of one million pounds for an Imperial Air Transport Company, and for the same amount to be paid as a subsidy by the British government over ten years. There should be two government directors on the Board.

Not surprisingly there were protests in the House of Commons when Sir Samuel Hoare spoke about the committee's report. For public money to be passed to a private company with shareholders was something which offended the basic principles of Labour members. Tom Johnston asked:

> Does not the Right Honourable gentleman consider it inadvisable that, of this committee of three persons, one should be his brother, another his brother's partner and the third the late partner of the Right Honourable gentleman himself?

Sir Samuel indignantly rejected the implied charge of nepotism. He admitted that none of the three had flying experience; their task had been to solve a business matter. Tom Johnston rose again to put Labour's point of view:

> For my part I trust that the government will retain civil aviation in their own hands, that they will run air transport for the benefit and well-being of the British people, and that private finance, private plunder and private capital will not be allowed to put its finger in this pie, as it has put its finger in telegraph and telephones.

Throughout 1923, while the Air Ministry negotiated with the interested parties, traffic was seen to be increasing but the companies were carrying under 500 passengers each week. Yet on those occasions when there was a much greater demand for space there were seldom enough aircraft to satisfy it. Handley Page Transport flew four services daily to and from the Continent, and Daimler Airway was able to match a German service to Croydon by operating one aircraft a week to Berlin, which returned the following day. Daimler was running single engine DH34s and inevitably the day came when one of these suffered an engine failure on a scheduled service.

This happened to their pilot, Ray Hinchliffe, who was able to put his machine down on the shore of a small Dutch island. This remarkable man, regarded as somewhat eccentric – he put on a bowler hat at the conclusion of each flight to walk from his aircraft into the Croydon terminal – wore a black or pink patch over one eye, the result of a wartime crash. With considerable enterprise he arranged for his DH34 to be transferred by a barge to Rotterdam. Daimler sent an aircraft with a spare engine to be installed. It was late afternoon by the time the necessary work had been completed and the two aircraft, urgently needed to fulfil the company's schedules, took off for England. It became dark over the Channel but only Hinchliffe's machine had navigation lights, so they landed at Lympne where traffic was very light. At dawn the following morning both pilots made the short hop back to Croydon.

In October continuous fog prevailed for an entire week during which only one flight was made. This was by O.P. Jones to Cologne.

Meanwhile The Instone Air Line submitted a 58 page proposal for the intended company, listing routes to be flown and the subsidy needed. A rival proposal was put forward by Szarvasy's group which was supported by Daimler and Handley Page. Agreement was finally reached on one scheme late in October and this was signed in December 1923.

The financial house guaranteed the first issue of £500,000 in shares for the purpose of buying out the assets of the companies. This included fifteen aircraft. Handley Page received the equivalent of £51,500 and the others lesser sums. There was however no rush to buy the shares, the City regarding air transport as a dangerous gamble. When Whitehall had put on sufficient

Capt. W. Hinchliffe. Cartoon by
Charles C. Dickson

pressure to find a financial house to underwrite the issue it was not long
before the shares were trading at a heavy discount. Only in the 1930s did the
public gain any confidence in the prospects of the new airline.

   The government nominated two directors to the Board. One of these
was Sir Herbert Hambling. The Instone Air Line was represented by Sir
Samuel Instone, Daimler Airway by Colonel Searle, Handley Page Transport
by Colonel Barnet-Lennard and Mr. Scott-Paine represented the British
Marine Air Navigation Company. The Secretary of State personally selected
a chairman. In his book *Empire of the Air* Sir Samuel Hoare wrote:

> As my chief object was to make flying a normal method of travelling it was essential to have
> as a chairman of the board someone who was a recognised expert on transport questions.
> Eric Geddes, formerly general manager of the North Eastern Railway . . . seemed to be the
> man best fitted for the post. It was true that he knew nothing of aviation . . . somewhat
> reluctantly he accepted the post, but only on condition that he brought with him Sir
> George Beharrel, the chartered accountant upon whom he always depended for financial
> advice.

Capt. O. P. Jones. Cartoon by
Charles C. Dickson

Geddes was Chairman of the Dunlop Rubber Company at the time and retained that position, acting as part-time Chairman of the new company, whose Managing Director was to be Colonel Searle. The latter accepted on the understanding that Woods Humphery should be General Manager. The Board met in the Dunlop office and employed the same firm of solicitors and auditors. The newly created airline had to be given a name and one suggestion was the British Air Transport Service. This was swiftly rejected when it was realised that the initials spelt BATS. Instead the name Imperial Airways was adopted and the contract was drawn up to inaugurate what was referred to as a "heavier than air" transport company. The contract also designated it "the chosen instrument" of the government.

Briefly the terms of the agreement were: firstly, that the Company was to enjoy a privileged position with regard to British subsidies for air transport in Europe; secondly, that the Company undertook to maintain services on the routes on which development work had been conducted by the companies previously operating; and thirdly, for ten years the Company was to be paid the subsidy by the government on a gradually diminishing basis and that in return for this sum of one million pounds Imperial Airways agreed to achieve a minimum amount of flying of one million miles a year over the first four years. The crews had to be British subjects and also to be members of the RAF Reserve. The aircraft were required to be of British manufacture and in the event of a national emergency the government had the right to take over all the Company's equipment.

With the winding up of the original airlines the first stage of the pioneering phase had been completed. Between August 1919 and March 1924, 34,605 passengers had been carried. Only five passengers and six crewmen had been killed. The Instone Air Line had operated without injury to any passenger. This was an astonishing achievement given the dubious reliability of the aircraft and engines in service at the time, the weather which affects northern Europe, and the very primitive facilities available to pilots for navigation and at the airports.

Although Imperial Airways had been formed in the autumn of 1923 it had not been registered as a company, nor had the executive come into being until the end of March 1924. No plans had been published. One possible reason may have been that in January the Labour party had formed

Handley Page 0/400 at Cricklewood, 1919.

a government for the first time, albeit dependent upon the support of the Liberals. Geddes and his Board may well have wondered whether their new political masters would permit a private monopoly company to exist.

Those investors who had been persuaded to buy shares in Imperial Airways cannot have been happy to learn that the new company had been prevented from commencing operations on 1st April 1924, owing to the refusal of the pilots to fly for the salary on offer. The delay in offering contracts to the pilots had caused them to believe that, as soon as they had ceased to receive the salaries paid by their former employers, they would be offered pay at a lower scale. The former Daimler pilots had received annual salaries of £1,000. Handley Page had paid £915 and Instone £827. The pilots were also aware that, as with other categories of staff, not all the former employees of the defunct airlines would be offered jobs.

Another grievance related to the appointment of Major Woods Humphery as General Manager. He had been a pilot during the war but his role in both Handley Page and later with Daimler had been in management. It was alleged that during his Handley Page days he had put pressure on pilots to fly against their better judgement and then put the blame upon them for damage to aircraft.

Late in March when the Board offered contracts to sixteen pilots the latter were outraged. Their salary was to be £100 a year and twopence per mile flown. At that time the aircraft averaged 86 mph in the cruise but bad weather often interrupted schedules and caused the cancellations of flights. When the Board were informed of the discontent engendered by their proposals a new formula was devised that was claimed to represent an average salary of between £750 and £850, depending upon seniority. This offer laid down the condition that the contract could be terminated at one day's notice. The pilots' request that flying pay should be based on time in the air rather than mileage was ignored.

The response of the pilots was to meet at Croydon and to form a union. The ground staff were not slow to follow their example. The Trades Union Congress gladly offered advice. On 1st April 1924, the day the Company should have started flying in their new livery, a delegation of pilots went to see the Labour Secretary of State for Air, Lord Thomson. They were accompanied by two well known trades union activists, Ben Tillett of the Dock Workers and Bob Williams of the Transport Workers. These two men were old allies who had tried in 1921 to bring out on strike all their members in support of the miners. Lord Thomson was told that Imperial Airways had been most unreasonable to delay so long in revealing their intended pay scale and its niggardly level. In addition they expressed their resentment at the hostile tactics of the Board in asking the RAF to release pilots to the Company.

For Imperial Airways Sir Eric Geddes and Colonel Searle informed Lord Thomson that the delays had been due to the time taken to obtain Certificates of Airworthiness for the aircraft which had been acquired from the old companies. The airline was now in a position to commence operations. They were prepared to pay between £755 and £855 if the pilots averaged two hours' flying a day.

By this time the dispute had become bitter. The pilots' union declared Woods Humphery to be lacking in the knowledge and experience to hold down the job of General Manager. He responded by serving a writ, filing suit for libel. The Air Ministry and other interested parties intervened. Sir Samuel Instone suggested that an Air Superintendent should be appointed, someone whose reputation was known and respected by the pilots, and upon whom they could depend to represent their views to the Board. This idea was adopted and Woods Humphery recommended Major H. G. Brackley for the post.

Brackley had been a junior clerk working for Reuters in 1914, but had served with distinction as a pilot during the war, being awarded the DSO and DSC. In 1919 he had joined Handley Page on whose behalf he attempted to win Lord Northcliffe's prize for the first successful crossing of the North Atlantic. Mechanical problems caused him to make a forced landing in Nova Scotia. Northcliffe offered another prize of £10,000 for the first flight to Cape Town, and Brackley piloted a Handley Page machine on that endeavour. This too ended with a forced landing, in the Sudan. Thereafter he had been a member of a British team assisting the Japanese naval air arm. He was back in England looking for a job when he was invited to apply for the post of Air Superintendent.

The terms of a contract acceptable to the pilots were agreed before Brackley had been appointed, but the former were reassured by his selection as their representative. The chief pilot, Franklyn Barnard, who had been the first pilot to be hired by the Instone brothers, wrote to Brackley in warm terms. The pilots' union quickly dissolved itself, most of its members unhappy that their grievances had been used as a political weapon by left wing activists.

The salary scale for the pilots was settled at a maximum of £880 per year with life insurance of £1,000. The contract was for a year, after which three months' notice could be given by either party. Although only sixteen pilots could be offered work at once the remaining twelve were promised that when a post was available they would be taken on.

With hindsight it seems that the Board of Imperial Airways blundered when they failed to explain to the pilots why they had waited to the last moment before revealing the terms of the contract which they proposed to offer. Not only did the Board have doubts about the first Labour

Franklyn Barnard and Sir Samuel Instone after the former's victory in the King's Cup Air Race in 1922. In 1924 Barnard became Chief Pilot of Imperial Airways. *Croydon Airport Society.*

government's intentions over the private monopoly company, but they were faced with the need to satisfy the expectations both of the shareholders and the government itself. Consequently an unworthy attempt had been made to economise at the expense of those upon whom the Board had to rely for the safe operation of their aircraft.

Brackley's satisfaction over his appointment as Air Superintendent was tempered by the knowledge that he had been placed in an invidious position. He was not a businessman nor a bureaucrat but had made his reputation as a brave and skilful airman. His post gave him access to Sir Eric Geddes and the Managing Director Colonel Searle, both of whom regarded him as a member of the management. At the same time the pilots depended upon him to support their representations to their employers.

Reviewing the history of the company after the first ten years *The Imperial Airways Gazette* avoided mention of this dispute and attributed the delay in the introduction of services to a "period of reconstruction". Brackley certainly found plenty to do. There were agreements to be

23

prepared for the initial sixteen pilots who had been engaged and interviews with other disgruntled men who could not immediately be employed by the new company. Woods Humphery had to be persuaded to withdraw his writ of libel. The combined aircraft of the former airlines had to be inspected and insured. Tons of freight were accumulating for shipment abroad.

It was decided that the pilots should fly on all the aircraft types and be prepared to operate on any route. The most perfunctory instruction was given and pilots who had never flown anything other than single engined machines were sent off, with occasional disastrous but not fatal results. Those pilots gave the most devoted service to Imperial Airways and even today their names will evoke memories among the older generation. They were Barnard, Bailey, Dismore, Hinchliffe, Horsey, O.P. Jones, McIntosh, Olley, Powell, Rogers, Robinson, Wilcockson, Youell and Wolley Dod.

Capt. Robinson. Cartoon by
Charles C. Dickson

CHAPTER THREE

The European Services 1924–28

Imperial Airways began operations with the combined fleets of its predecessors but only thirteen of these aircraft were air worthy. Seven DH34 single-engined biplanes had been owned by Instone and Daimler. Each of these had originally cost £10,000 but were valued at half that figure. With seats for eight passengers they were flown by one pilot whilst a wireless operator also acted as an engineer. Colonel Searle was keen to replace these machines with multi-engined aircraft that would be more economical to operate and carry more passengers. On the DH34s operated by Daimler, cabin boys were employed. Aged about fourteen, they were trained by the Savoy Hotel and dressed like page boys. On the aircraft there were no essential duties for them to perform as neither food nor drinks were served.

Originally the property of Handley Page Transport and built by their founder, three twin-engined W8B biplanes were valued at £10,000. They had seats for fourteen passengers and a crew of two, a pilot and wireless operator. This was the first post-war civil air transport. Two Supermarine Sea Eagle amphibians were twin-engined biplanes which had belonged to the British Marine Air Navigation Company. Crewed by a pilot and wireless operator they carried six passengers. Imperial Airways insured each of these for £5,000.

Finally there were two aircraft which Instone had used to carry passengers but which Imperial Airways used solely for freight; a Vickers Vimy twin-engined biplane which had originally cost £10,000 and could carry ten passengers was insured for £4,500, and a Vickers Vulcan single-engined biplane, which had also carried ten passengers, was valued at £3,800.

The seating on all these aircraft comprised wicker chairs in a narrow cabin. No provision was made for heating, nor was there any insulation from the noise of the engines. There was a lavatory on some of them, but if the passengers had the stomach for eating in often turbulent conditions they would have had to purchase a lunch box at the airport canteen before departure.

Imperial Airways was able to start its services on 26th April 1924, when a daily flight to Paris was instituted. On 3rd May services to Brussels and Cologne got under way with an additional flight to Cologne via Ostend. On 2nd June an aircraft departed for Berlin via Amsterdam and Hanover. This service was operated three times a week in each direction on alternative days

This Handley Page W.8.b acquired by Imperial Airways from Handley Page Transport was in service until 1932. It was the first post-war aircraft to enter service. *Croydon Airport Society.*

by both Imperial Airways and the German airline Aero Lloyd. On 17th June flights to Zurich via Paris and Basle inaugurated a thrice weekly service. It was hoped to run daily flights to Guernsey with the Sea Eagles, but weather and the sea conditions frequently caused cancellations.

During 1924 Imperial Airways decided that if more passengers were to be attracted a higher standard of comfort and reliability was needed. A policy of using larger, more commodious three-engined airliners was adopted. In November the first of these was commissioned. Built by Handley Page, a Rolls-Royce Eagle engine was fitted to the nose of the machine, with a Puma engine on either side of the fuselage. Called the W8F this aircraft was flown with two pilots and carried twelve passengers,

In a handbook, 'Points for the Passenger', which Imperial Airways issued at that time it was stated: "The passenger cabin is entirely enclosed. The windows on either side can be opened or closed at will. There is no more need for special clothing than there is on a railway journey." The term 'special clothing' was a reference to the helmet, goggles, flying suit, boots and so forth that were still being worn by the pilots who sat exposed to the elements behind a windscreen. This was frequently spattered by oil blown back from the engine immediately in front of it.

There was a general election in 1924 and Ramsay MacDonald's short lived Labour government, deserted by their Liberal allies, left office. Sir

Supermarine Sea Eagle, the first type of flying boat used by Imperial Airways, formerly operated by the British Marine Navigation Company, 1924. *Mr E. B. Morgan.*

Samuel Hoare resumed his former post as Secretary of State for Air. He was to perform this task until 1929 when Lord Thomson of Cardington, an enthusiast for the nation's doomed airship programme, again replaced him. Hoare clearly relished his appointment and made it his practice to travel by air to perform his duties. He incurred the disapproval of the Prime Minister, Stanley Baldwin, who regarded flying as a dangerous activity. So too did the London *Morning Post* which declared: "The place for ministers is in White-hall and when ministers travel they should keep to the established methods of transport."

Many years later Hoare wrote:

> What better in those days when every civil flight was regarded as a foolhardy adventure than for the Secretary of State for Air to start using aeroplanes for his official journeys … the idea that he should normally use air transport for visits and inspections was entirely new and in the minds of many people, mad … my flights they regarded as objectionable stunts. In actual fact they were the very opposite of stunts. Their whole object was to prove that flying was not a stunt, and I did everything possible to make them as humdrum and unsensational as possible.

Even so the year 1924 ended in tragedy when a DH34 bound for Paris on Christmas Eve crashed on take-off from Croydon, killing all eight persons on board. The subsequent inquiry established that the pilot had appeared to have difficulty getting the airliner off the rising ground towards Purley. The chief pilot of Imperial Airways, Franklyn Barnard, attributed the crash to the layout of the aerodrome which made take-off difficult when the wind was from the south-west. Another pilot, Lieutenant-Colonel Minchin, declared that the 1600 yards available was not long enough and that the solution lay in the removal of some buildings.

Air Superintendent Brackley agreed that in bad weather Croydon could be unsafe for flying. Allegations that nothing had been done to rectify an engine fault reported by the one-eyed pilot, Hinchliffe, on a previous flight, were denied. "The facts did not call for more than a ground run of the engine," was the Company's response. This had been done and the fault had not been reproduced. The previous year a report on the layout of Croydon had recommended major changes, the principal one being that aircraft should not be required to take off uphill. After this disaster it was decided to utilise 150 more acres of land to the west of the aerodrome and to redesign the entire layout.

Lt. Col. F. Minchin. Cartoon by
Charles C. Dickson

As Croydon became busier, a letter to *The Times* suggested that the London Underground system should be extended to reach it. This idea was never adopted. Complaints over aircraft noise grew, with local residents increasingly resentful of the testing of engines in the early hours of the morning. One householder attempted unsuccessfully to obtain a summons against Imperial Airways. Subsequently he tried to persuade local residents to band together to decide upon some effective action. Complaints continued throughout the following year, the local newspaper printing one letter which concluded: "I cannot see the slightest reason why a commercial company subsidised with public money should be allowed to spoil the whole district and make sleep impossible for hundreds of people."

Before the inauguration of Imperial Airways, plans for a service to Prague had been drawn up by The Instone Air Line. It had never been possible to start this owing to obstruction by the Germans and the French. The former were resentful of conditions imposed by the Allies on their air transport. Moreover there was no advantage to them whatever in a British service landing on their territory *en route* to Prague. On the contrary it represented competition. The French, still bristling over the non-payment of war reparations, were further incensed by Germany's confiscation of ten of their civil aircraft. French plans also included a service to Prague. Negotiations proceeded fruitlessly for years, the keenness of Imperial Airways and the British government arising from their appreciation that Prague's geographical position in Europe made the city an ideal staging post on the projected route to the Middle and Far East. When an air traffic agreement with the Germans was finally achieved the Company could no longer agree terms with the Czechs and the plans to fly to Prague were shelved.

In April 1925 readers of London's serious newspapers learnt that the general manager of Imperial Airways had resigned. Colonel Searle's priorities were safety first and then the highest level of efficiency. With a small number of aeroplanes 800,000 miles had been flown in the first year of the Company's existence, but at a considerable loss. Sir Eric Geddes and his Board were businessmen. Their principal concern was to earn revenue and to reduce expenses. Searle's departure from the Board resulted in the promotion of Woods Humphery to fill his former colleague's post of Managing Director.

Six serviceable aircraft were required to operate the airline's daily schedule, but in the summer of 1925 it was not always possible to achieve this number. On one Friday in August there was no flying due to the unavailability of aircraft. This situation attracted a lot of adverse comment both from Members of Parliament and in the national press. Early in September it was announced that three new Handley Page W10s had been ordered. At the end of the year Sir Samuel Hoare claimed that 51 per cent of the passengers flown to and from Croydon had travelled by Imperial Airways. Nevertheless it was obvious that the Company was not keeping up with the competition. Months earlier Sabena had opened a regular service through Spain, the Sahara and Nigeria to Lake Chad and Leopoldville in the Belgian Congo. In August the Germans had flown into Croydon an all-metal three-engined Junkers G24 monoplane.

The Aeroplane published an article of venomous sarcasm:

> One gathers that the present numerical strength of Imperial Airways is according to plan and it is not due to the short-sighted policy of the directorate. To qualify for the subsidy their aircraft have to complete one million miles a year. This is all planned so that on the last day of the financial year the millionth mile is achieved. When, as was the case last year, they run it too fine and are short of the million, they make excuses to the Air Ministry that a strike or a pestilence or a famine held them up. If they find they are covering too much mileage a machine can be taken off the service and passengers kept waiting and told no machine is available. That is why the French are reaping a reward in the shape of passengers. Imperial Airways have sold out to the French for a mess of pottage (or at any rate a mess) the birthright of British civil aviation, built up so well in the past by the pioneer companies.

The editor of *The Aeroplane* was C. G. Grey who had also founded this weekly magazine in 1911. Before that he had been a founder member of the Royal Aeronautical Society in 1909. At a time when the newspapers only showed an interest in civil aviation when there were record-breaking attempts or spectacular crashes *The Aeroplane* was the obvious source for anyone anxious to glean authoritative information. C. G. Grey was on good terms with Woods Humphery, for whom he had a very high regard, but he did not let his friendship temper his criticism of the Company, particularly its neglect of good public relations.

When four Handley Page W10s in the new livery of a dark blue fuselage were accepted by Imperial Airways in the spring of 1926 Grey was not impressed:

> The engines are unhoused and uncowled. Control cables and pulleys and a species of sheet metal stick out all over the machine in wartime fashion. Even the cabin chairs are of the old straight-up short-backed type of inflammable wicker, which provides about the least comfortable seating imaginable.

Handley Page also completed the prototype of the three-engined W9A. A ceremony at Croydon's Aerodrome Hotel that was attended by Sir Eric Geddes celebrated these events and nobody was sufficiently unkind to observe that the W10 had been designed seven years earlier. During the year the entry into service of the W9 was followed by two of the Armstrong Whitworth Argosies which had also been ordered. These were to prove to be the workhorses of the Company for almost a decade. They were also the first Imperial Airways airliners to carry a steward and possess a buffet. Each passenger sat by a window that could be opened. But these aircraft were

Imperial Airways. G-EBMS. Handley Page. W.10. 1926.

extremely draughty and notorious for their continuous vibration. Moisture dripped from roof panels on to passengers seated beneath them.

Like the Handley Page W10s delivered earlier in the year, the first three Argosies had their fuselages painted a deep blue and silver-doped wings. Both this livery and the 'City' names were a continuation of The Instone Air Line practice, but within a year aircraft were being repainted all-silver. Naming continued but from 1931, when the 'Scipio' class were delivered, new aircraft were given names of classical origin. As Appendix 1 shows, the smaller or older aircraft, used mainly on charter work or the Channel Islands route, were never named.

The pilots were issued dark blue uniforms in the style of the Merchant Navy. Some of them may have welcomed this but viewed with mixed feelings a new regulation issuing from the Air Ministry. This required airliners carrying ten or more passengers a distance of over 100 miles to include among the crew one who possessed a navigator's licence. Imperial Airways ordered their pilots to obtain this qualification and, following maritime practice, expected them to purchase sextants at their own expense. The Company also left it to the pilots to study in their own free time from whatever instructional material they could find. To their great credit they all passed the examination.

The General Strike, called by the Trades Union Congress in 1926 in support of the miners, paralysed the railways and many transport companies but provided Imperial Airways with a huge demand for seats. The service to

31

Paris was tripled and would-be passengers at le Bourget, unable to obtain an air booking, tried to force their way on to airliners. Every aircraft travelled full and although the TUC called off the strike after eight days, business continued to be good.

In the course of 1926 Air Superintendent Brackley was shocked to be given notice of dismissal on the grounds that he had not performed his duties at the standard expected of him. Brackley was aware that it was not the custom of Sir Eric Geddes to deal directly with his employees and he normally only did so in the presence of Woods Humphery. It was said of the Chairman that he regarded his pilots as no different from the engine drivers in his employ in the days when he ran the North Eastern Railway. Brackley therefore spoke to Woods Humphery to enquire in what respects he had failed in his duties.

The Managing Director chose to reply in a letter, dated 25th October 1926:

> With reference to our conversation regarding your position in the Company, I have given very careful thought to this matter and whilst I do not doubt that you have done your best to carry out your duties satisfactorily, and much as I regret being obliged to say so, you have not rendered the standard of service which the Company expects from a highly salaried official in your position. You pressed me to say where you have failed – it is in the case of a post such as yours almost impossible to do so without inviting controversy, but amongst other things, your failings have included a lack of foresight, tact and leadership; also a failure to take advantage of the talent which undoubtedly exists among our pilots to provide the Company with information, well considered and balanced views on matters of importance to the technical development of the Company, or to assist me in the question of policy which should arise from time to time in your daily work. I have therefore with regret had to tell you and the Board that it is not in the best interests of the Company for you to continue in your post.

Woods Humphery concluded by stating that the Board would reconsider their decision if a higher standard of service was rendered within the following three months. In the event Brackley retained the post of Air Superintendent until 1939.

In her biography of her husband published in 1952 Mrs. Brackley wrote of the relationship between the two men:

> Both were young, enthusiastic, ambitious, self-made and single minded in their devotion to their ideal of work: temperamentally poles apart with a new science and unexplored business to pursue. I can only wonder more and more, despite all the difficulties, disillusionments and misunderstandings of others, mostly, and at times between themselves, what an amazing record they both achieved.

Brackley's competence as a practical airman did not go unappreciated by those who flew and serviced the aircraft. He held a commercial pilot's licence and tested every new aircraft before Imperial Airways introduced it into service. Lloyd Ifould, the station engineer at le Bourget recalled the occasion when the Air Superintendent arrived early one morning at the airport. Overnight Ifould had changed an engine on an aircraft that was scheduled to depart at 9 a.m. He informed Brackley that there was bound to be a delay to the service because when the pilot arrived he would have to conduct an air test. Without a moment's hesitation the Air Superintendent replied, "Start the engines. I will test her myself."

In December 1926 three of the new three-engined DH66 Hercules were delivered in quick succession. This type had been designed, built, tested and handed over to the Company in the course of one year. No trial flights were carried out on Imperial Airways routes and the aircraft were put into service with a minimum of delay. Wolley Dod, chief pilot of the eastern route, flew the first Hercules to Cairo to be used for the mail service to Baghdad. He was

C. F. Wolley Dod conducted many survey flights in Africa and along the eastern routes. He was killed in 1937 when a DH86 crashed in a storm near Cologne.
R. A. R. Wilson

Capt. Horsey. Cartoon by Charles
C. Dickson in the Pilots' room of
the Aerodrome Hotel, Croydon.

followed by Hinchliffe in the second and Barnard in the third Hercules. This airliner, like the Argosy, had an open cockpit because the pilots hated to be enclosed in what O. P. Jones called a "bloody chicken coop." The wartime fliers liked to feel the airstream on their cheeks, believing that it enabled them to assess the drift from the wind as they made their approach to land.

As the year came to an end Imperial Airways announced that it had carried 16,652 passengers in and out of England. An M.P. who asked how this figure compared with German airline traffic was told that the information was not available. C. G. Grey published an apt comment. "Comparing Imperial Airways traffic with German traffic is like comparing traffic out of Penzance railway station with that of Manchester." It was a fact that Luft Hansa was already the biggest air traffic combine in the world.

The last of the Company's bomber conversions had been withdrawn. The fleet now comprised six twin-engined and five three-engined machines. The accident record over the previous twelve months had been an unhappy one. A Sea Eagle was destroyed at its moorings off Guernsey when it was rammed and cut in half by a steamer. A Handley Page W8B, piloted by H. Horsey and flying beneath the clouds from Amsterdam to Croydon, had an engine failure. Horsey managed to reach the English coast but realised that he could not clear the tops of the cliffs so made a forced landing on the beach near Dover. In October one of the new Handley Page W10s, flown by Dismore, had to ditch in the English Channel. Fortunately there were no fatalities in any of these incidents.

In January 1927 heavy snow falling on a huge canvas hangar at Croydon caused its collapse, severely damaging the Company's W10 which was inside. The snow did not discourage McIntosh from taking off for Paris with passengers whom he thought would otherwise grumble about being delayed. But the weather *en route* was so bad that he found a field to land in, and completed the flight the following day.

In April Imperial Airways introduced, at a surcharge of one guinea, a luxury service to Paris which they called the 'Silver Wing' flight. The passengers were brought from London to Croydon in a coach, attended by a uniformed page boy. The Argosy was equipped with a buffet and the fare was £6.6s one way. But that summer the airline was plagued by more problems. A Handley Page W9 crashed at Biggin Hill, having run out of fuel much sooner than expected. An aircraft shortage was compounded by a shortage of such parts as propeller blades.

Several tragedies reduced the number of available pilots. Inspired by Lindbergh's successful trans-Atlantic flight to Paris, and bored by routine cross-Channel trips, Minchin decided to make an attempt upon an east to west crossing in the company of a sponsor, the Princess Löwenstein Wertheim. As they headed west their Fokker's navigation lights were seen by crewmen on an oil tanker. Some time later a wheel of their aircraft was washed up on the shores of Iceland.

More poignant was the death of R. H. Hinchliffe, who had shown himself to be such a dedicated and competent pilot. He had joined Imperial Airways from Daimler, but the stiffer medical requirements that came into force caused him to lose his job. With a wife and three children to support he had anxiously sought a sponsor for a record-breaking flight, which would bring him a financial reward. He found one in the person of the Hon. Elsie Mackay, who was the daughter of the chairman of the P & O Line, Lord Inchcape. Hinchliffe chose for their trans-Atlantic venture a Stinson Detroiter which he imported from the United States. This young couple also disappeared without trace over the huge breadth of the ocean.* Chief pilot Barnard was killed while testing a private aeroplane in preparation for the King's Cup Air Race.

*See Appendix 5 for a more detailed account of the tragedy.

There is much truth in the adage that there are old pilots and bold pilots but no old bold pilots. William Armstrong had been a pilot in the Royal Flying Corps before joining AT & T. In his autobiography, *Pioneer Pilot*, he admitted how fearful he was of the weather with which he had frequently to contend between Croydon and Paris. Without any of the aids available to his successors he recalled flying a few feet above the waves of the English Channel, unable to distinguish between the water and the mist, the sudden glimpse of a church spire looming above the level of his aircraft, as he tried to follow a road or a railway line. Small wonder that flying was universally thought of as a most dangerous form of travel. It was a common belief shared by many pilots as well as the public.

When AT & T went into liquidation Armstrong decided to qualify as a pharmacist and did so, but after the inception of Imperial Airways he was encouraged to apply for a job by Brackley. Suspicious at first that he was being hired as a strike breaker, he waited until the pilots and management agreed terms before accepting a contract. He was not a man who took unnecessary risks. The DH34 airliner which he had flown may have had only the one Napier Lion engine but it had been tried, tested and found reliable in wartime service. He was less impressed by the Handley Page W8B because it would not hold its height with one of the Eagle engines out of action. He had the same complaint to make about the Armstrong Whitworth Argosy when only two of its three Jaguar engines were working properly. At least the pilot had time to look for somewhere to land. He was a careful pilot and finally retired as a senior captain in BOAC after the second World War.

During the winter of 1927 Imperial Airways introduced a second class fare on the Paris route, on flights scheduled at the less popular times. No bar service was provided and no steward was on hand to attend the passengers. These promotions paid off and during the year 70 per cent of all the passengers on the Paris route in both directions, were carried by the Company. Moreover the third annual accounts of Imperial Airways disclosed a profit for the first time, the modest sum of £11,000.

In February 1928 bad weather prevailed and three Argosies parked in the open at Croydon were damaged in a squall. This followed the crash of an old Vickers Vulcan on a test flight, a type which had first entered service in 1922, when The Instone Air Line had bought three. Imperial Airways aimed to operate fifteen aircraft but was sometimes obliged to re-commission older machines to minimise delays and cancellation of services.

Run in competition with heavily subsidised foreign airlines these services lost money. Gradually Imperial Airways reduced their schedules, abandoning the Amsterdam route to KLM and allowing Luft Hansa to carry passengers to Berlin. In 1928 Sir Samuel Hoare agreed that the Company's principal task should be to concentrate its efforts on routes to the Empire.

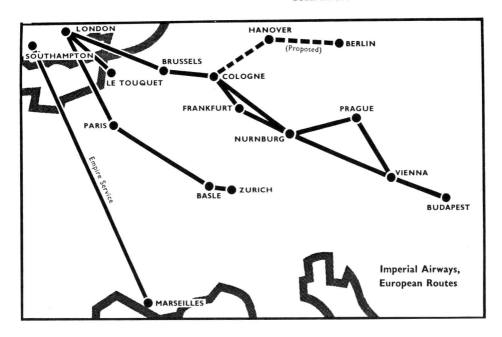

On the other hand no encouragement was given by the government to any other British company to develop services to Europe. As the 'chosen instrument' only Imperial Airways was entitled to a subsidy. As early as 1919 the Blackburn Aircraft Company had wanted to operate a service to Denmark and Sweden. Their plans fell through because the government refused to provide Customs facilities at an aerodrome in the north of England. Obviously Lympne in Kent, which did offer a Customs service, added too great a distance between the north of England and the Scandinavian countries to make the venture an economic proposition. In 1928 the Hull Chamber of Commerce again debated making their city part of an air network to northern Europe, but received no encouragement.

Imperial Airways was finding the million pound subsidy to be insufficient, given their obligation to pay a dividend to the shareholders. Consequently they allowed foreign airlines to pick up business from British concerns and acted as ticket agents for the former. Yet at the conclusion of four years of operations the Company was able to pay a dividend, the Board being determined to honour the prospectus and agreement under which the capital had been raised. "Under existing conditions in this small island," Sir Eric Geddes told the shareholders, "our future lies in long distance Empire routes."

With the approval of Sir Sefton Brancker, Director of Civil Aviation, the Company invited the British aircraft manufacturers to tender for airliners seating up to 40 passengers, both with three engines and four. Price quotations were to be for a minimum of three and a maximum of six aircraft. The emphasis was on high payload, low cost of operation and a stalling speed not above 52 mph. Many aerodromes were small. There was a response from five firms and Handley Page was awarded the contract. Their design was for the large four-engined HP42 which was eventually put into service in 1931.

Meanwhile a charter department was formed by the Company to seek out unscheduled business, rather than leaving the initiative to individuals and business. Gordon Olley was appointed to manage this venture and to carry out the flying. The Company's plan to introduce a combined air and rail transport system for cargo failed to receive the approval of the Air Ministry. One reason perhaps for the failure, in public relations terms, of a publicity stunt involving the simultaneous departure from London of an airliner and the *Flying Scotsman*, both bound for Edinburgh.

The pilot was Gordon Olley and he was accompanied in the cockpit by an engine driver. Another pilot travelled on the footplate of the locomotive. Journalists had been invited to sample both means of transport in order to contrast the advantages of each. The *Flying Scotsman* had only recently inaugurated a non-stop service to Edinburgh but Olley planned to land twice on route to refuel. He proposed also to fly directly above the train as it crossed the Royal Border bridge over the River Tweed at Berwick, and photographers were in position to take pictures of this scene.

Unfortunately Olley tracked the wrong train which passed over the bridge five minutes after the *Flying Scotsman* had already done so. His engine driver companion realised the error when this train stopped at the station. Olley increased speed, fearful of arriving late at his destination. After landing the passengers were hastily despatched to greet those who had travelled on the *Flying Scotsman*, but arrived at the railway station to find them already on the platform. This planned event did nothing to encourage the concept of a domestic air service.

CHAPTER FOUR

Flights to India 1921–32

Political and technical difficulties made it impossible for Imperial Airways to carry passengers over the entire distance to India until the 1930s, but ground studies for a future route network were constantly in progress. In this respect the pioneering flights of the RAF and others proved very useful. In 1918 Major-General Salmond flew from Cairo to Delhi in under 48 hours airborne time. The following year the brothers Ross and Keith Smith flew from England to Australia via India in a successful bid to complete the journey in under thirty days. Dr. John Ball, the Director of Desert Survey to the Egyptian government, prepared maps of the trackless wasteland east of Palestine as far as the Euphrates valley over which aircraft would have to fly.

In 1921 Sir Hugh Trenchard, Chief of the Air Staff, and Winston Churchill, whose portfolio combined Colonies and Air, attended a meeting of officers commanding British units in the Middle East. Held in Cairo, the purpose of the meeting was to inaugurate the Middle East Department of the Colonial Office. Of great importance however was their decision to give control of the area to Royal Air Force squadrons rather than to the Army. This not only allowed great economies in the deployment of manpower, but also provided very useful training for the airmen.

To assist navigation over the almost featureless waste it was decided to plough a broad track along the ground over the route selected by Dr. Ball. Convoys of vehicles set out from both Amman and Baghdad, making use of tractors where necessary, to ensure that the track was visible from the air. About every twenty miles along the route a circle was ploughed around ground suitable for a forced landing. Inside the circle was a number or a letter, depending upon whether the landing ground was east or west of El Jid. This was the boundary separating the responsibilities of the Jordan and Iraq Commands. Large arrows were also cut into the sand pointing in the direction of the nearest landing ground. Great care was taken to keep the track in good order, with pilots reporting when drifting sand obscured any parts of it. All flights were carried out in daylight by two aircraft travelling together.

Undoubtedly those who flew above the desert furrow so painstakingly prepared must have been grateful for it, but in 1922 Sir Sefton Brancker expressed some reservations when he made a report to the Air Ministry:

The present means of navigation is to my mind most derogatory to the training of pilots in navigation. I realise the necessity of this system at present as a safeguard against a pilot and machine being lost in the desert, and its occupants dying of thirst. I feel very strongly that a really efficient system of wireless telegraphy should be established on this route as soon as possible. The present system is likely to spoil a good pilot.

Even so the airmen faced dangers. There were many forced landings when their water-cooled engines gave trouble. Spare parts had to be flown to the emergency landing grounds and engine changes made in extremely high temperatures during summer. It required sixty cans of petrol to put fuel sufficient for a five hour flight into the tanks of two aircraft, quite apart from cans of benzol and drums of oil. Someone whose only identification is the letters J.O.P.E. wrote a ballad about this route. He described the sight of the blackened wreckage of an aircraft whose pilot had lost his way in bad weather. The last verse went as follows:

We have placed at your disposal all the acts we learnt in war,
And for nine pence you can purchase if you're wise,
The same efficient service of the man who yesterday
Chased the black-crossed birds of war from out your skies.
The letters that they carry from Cairo to the East
Bear little slips of paper coloured blue,
And the loss of crew and pilot doesn't matter in the least
If the mail bags see the desert journey through.

Government mail was carried over this route but when the public were also offered the service it was some time before it was fully patronised. The mail was sent out to Cairo by surface transport but as there was usually a period of two weeks between flights there was no great saving in time.

In November 1924 Sir Sefton Brancker accompanied Alan Cobham and a mechanic on a survey flight to Rangoon. The Air Ministry made a small contribution to the costs but the greater part was shared between Imperial Airways, The Society of British Aircraft Constructors, The Anglo-Persian Oil Company and the de Havilland Aircraft Company. They flew out in a single-engine DH50 with stops in Paris, Cologne, Berlin, Warsaw, Bucharest, Constantinople, Ankara, Aleppo, Baghdad, Esfahan, Quetta, Kerman, Lahore, Delhi, Allahabad, Calcutta and Akyab. The trip was not without incident.

Brancker cut his head on the roof of the cabin. At one stop the aircraft tipped over on to its nose. In Calcutta he developed pneumonia but continued on to Rangoon. Homeward bound they made a forced landing in a snowstorm in Germany. Assisted by the local German police they dismantled the aircraft and had it transported in three trucks to an aerodrome near Stuttgart. The following morning they re-assembled the machine and continued to Strasbourg, returning to Croydon in March 1925.

Brancker had realised some of the difficulties that an airline would face over such distances, from handling passengers after a forced landing, to the adjustment of loads to compensate for either very high ground temperatures or the rarefied air of aerodromes situated several thousand feet above sea level. In such circumstances the take-off performance of an aircraft is degraded. However the principal reason for the flight was to find out how co-operative the governments of those countries over which they would fly, or upon whose territory they proposed to land, would be. Whereas some might offer a subsidy, in order to develop trade links with their neighbours, others might refuse passage altogether.

Brancker received satisfactory assurances from the Turks. The Iraqis did not wish to incur any expense to themselves, but were willing to co-operate in the establishment of an aerodrome at Rutbah Wells, which lay midway between the eastern coast of the Mediterranean and Baghdad. Less helpful was the attitude of the government of Persia, as Iran was then called. There was no question of a subsidy, and favourable treatment was being accorded to the Junkers Company of Germany. The latter planned to organise a flying school in Teheran and to operate three times a week to Baghdad with a subsidy from the Persians. The policy of the German government, designed to compensate for the constraints placed upon it by the Allies, was to develop useful contacts in the East. This included Persia, Russia and China. It was also intended to give useful experience to the officers of their much reduced army.

Brancker observed that in India no-one in authority knew anything about aviation. The Indian government was prepared to co-operate only on two routes, from Bombay to Karachi and from Calcutta to Rangoon, both of which served the interests of their military advisers. Even so he was not dismayed by any of these reactions. His report on the feasibility and prospects of the route to India expressed optimism. Brancker correctly suspected that it would take longer to establish routes across Europe and the Mediterranean.

The British government's reaction to Brancker's report was to replace the existing RAF mail service in the Middle East by a passenger and mail service to be operated by Imperial Airways. It agreed to pay a subsidy of £93,600 in the first year of operation and to take responsibility for aerodromes, hangars and accommodation in territory under its control. The money would be taken from savings in the Defence Vote. In August 1925 Imperial Airways and the Air Ministry jointly conducted an aerial survey of the route, Minchin landing at every possible aerodrome which it might be necessary to use. Shortly afterwards a ground survey was carried out in which the pilot Wolley Dod participated.

The inhospitable nature of the terrain induced Imperial Airways to employ three-engined aircraft. This decision would put up the costs worked out for a weekly service so the planned schedule was revised to a service each fortnight. In September 1925 Brancker went to Cairo to conduct secret negotiations with the Persians. They agreed to allow an air route along the southern shore of the territory, that is along the northern shore of the Gulf. He flew on to Persia to complete the arrangements before returning to England.

On the strength of Persian assurances Imperial Airways ordered four de Havilland DH66 Hercules aircraft. These had an endurance of five hours and an operational ceiling of 13,000 feet, sufficient to clear the high ground on the way to India. To allow sufficient space for mail bags seven of the fourteen passenger seats were removed. The Company took delivery of the first of these machines in November 1926, and the following month *City of Cairo* left Croydon for Egypt. In January 1927 this aircraft made the first scheduled mail flight to Baghdad.

As Sir Sefton Brancker had foreseen, obstacles were placed in the way of flights by Imperial Airways over Europe. The French would not permit Imperial Airways to fly beyond Paris to Genoa, because the Company had refused to participate in a joint flying-boat service from Marseille to the eastern Mediterranean. This made it necessary to send passengers in an Argosy to Paris, and onward by rail to Marseille. At this port they could

DH66 Hercules, used by Imperial Airways on the route to India.

embark on a steamship of the P & O Line which plied every week to Alexandria.

Meanwhile studies were continuing for a suitable route across the Mediterranean and an appropriate aircraft to operate it. During 1926 RAF flying-boats had flown east as far as Cyprus to obtain experience of this type of aircraft. In the House of Commons the Hon. F.E. Guest called for a civil flying-boat to be developed and suggested Southampton as a suitable base for them. Major Brackley was also a flying-boat enthusiast and he urged Brancker to persuade the Air Ministry to order a civil version. As a result of these representations the Air Ministry placed an order with Short for a number of Calcutta flying-boats. Developed from the military Short Singapore, the three-engined Calcutta was unique among flying-boats at the time in being constructed almost entirely in metal, and became the first civil flying-boat capable of crossing the Mediterranean.

In the summer of 1928 one was tried out on the route between Southampton and Guernsey. Then the corporations of Liverpool and Belfast subsidised Imperial Airways to run a second Calcutta between their cities. Although this flying-boat cruised at under 100 mph, and the pilots occupied an open cockpit, a well upholstered cabin accommodated fifteen passengers, who were permitted to smoke, as the fuel was carried in the wings. The seats were fitted with pneumatic cushions which could be used as life-rafts in an emergency. Costing £18,000 each the Calcuttas had been delivered to the Company as a capital grant. Even if the passengers for destinations beyond Cairo could not be flown across France, it remained the plan that they should be flown across the Mediterranean by the Calcuttas, when these had proved themselves nearer home.

Heliopolis, on the outskirts of Cairo, was the designated Egyptian terminal and whenever possible RAF aerodromes were used. Accommodation for passengers along the route was anything but luxurious. At Rutbah Wells a landing ground was prepared in the desert and a fort constructed at the oasis nearby. Inside the fort a rest house was built for the permanent staff. It was supplied by trucks which travelled along the desert tracks from Baghdad, 240 miles away. There was a plentiful supply of fresh water from an artesian well but the architect had neglected to provide fireplaces, forgetting that a bitterly cold temperature prevailed in winter over areas which lay 3,000 or more feet above sea level.

The Hercules airliners which were employed on this route could maintain contact with ground wireless stations and were not dependent upon the ploughed furrow for navigation. In case of a forced landing, desert rations and an emergency water supply were carried. In addition emergency fuel dumps were laid down. The petrol cans, although not their contents, so appealed to the desert nomads that they had to be locked underground. In case a pilot had forgotten to carry the key, it was the idea of Woods Humphery that these should be identical to those required to open the aircraft cabin doors. Although the nomads took a delight at using the locks for target practice and scorpions hid between the fuel tanks and the surface of the desert, these measures proved satisfactory on the occasions when the fuel was required.

Sir Samuel Hoare, accompanied by his wife, travelled in *City of Delhi* on the first Hercules flight to India, in January 1927. Stops were made at Dijon, Marseille, Pisa, Naples, Homs, Benghazi, Sollum, Aboukir, Bushire, Lingeh, Jask, Pasni, Karachi and Jodhpur before they landed at Delhi, after a flying time of 65 hours "We were both," he wrote, "intent upon proving to a doubting world that flying was a normal and dependable way of travelling, for women as well as for men, and no longer an adventure that only men could undertake." He believed that the publicity could only be of benefit to British aviation. Certainly their safe arrival in India was rewarded by a message of congratulation from King George V and the Khan of Kulat offered him "a carpet as a symbol of your conquest of the air." In England Prime Minister Stanley Baldwin, commenting upon the Secretary of State's journey and that of another minister, Leo Amery, skiing in the Alps, remarked that he felt "like a circus manager whose performing fleas have escaped."

Sir Samuel and his wife returned to England in February, completing the journey from Paris by sea because of fog. *The Aeroplane* congratulated them for their courage. "Few in their circumstances … would risk their lives merely because they considered their duty was to set a good example to the travelling public."

The air service operated without a serious incident until December 1927 when *City of Cairo* reported fuel so low that the pilot intended to put down between landing Grounds Four and Five and asked for fuel to be sent out. The vehicle carrying this failed to find the machine and two days passed before another Hercules spotted the lost airliner fifty miles south of the reported position. The pilot landed close by and found that the crew and passengers had been courteously treated by the local tribesmen. After fuel had been flown in, both aircraft were able to reach Baghdad that night.

Yet once again political intrigue frustrated the Imperial Airways schedule. The Russian government had not forgotten that the British had used Persia as a base to help the White Russians after the Bolsheviks had seized power. Pressure from this source, and the refusal of the British government to allow the Junkers Company to operate a scheduled service from Teheran to Baghdad brought about a revocation by the Persians of their agreement with Imperial Airways. After prolonged negotiations they offered a corridor for international flights across central Persia. This could

Short Calcutta. One of five which were in service between 1928 and 1937. *Robert Jackson.*

not be accepted because of the lack of aerodromes and the mountainous terrain along that route. Finally the Persians offered to negotiate with Imperial Airways alone, excluding an official British representation. In May 1928, after spending seven weeks in Teheran, during which he was received by the Prime Minister on only four occasions, Woods Humphery obtained an agreement for the airline to fly the route along the southern shore of Persia. As the aerodromes were not ready Imperial Airways was obliged to terminate its east bound service at Basra in Iraq.

The Foreign Office considered that there was a real possibility that the Russians might send troops to occupy the southern shore of Persia. As a result the Air Ministry and Imperial Airways jointly surveyed the Arabian side of the Persian Gulf, the territory of the Trucial Sheikhdoms, including places now well known to many international travellers: Abu Dhabi, Dubai, Sharjah and Muscat. This was a wise decision because the Persians laid down a new corridor for aircraft across mountains over 10,000 feet high, salt deserts which turned into quagmires in winter and areas without roads, telegraph or wireless stations.

Nearer home, the Channel Islands service was discontinued in May 1929 and the Calcuttas were transferred to the Mediterranean route. The Argosy was used to carry mail from London across Europe but no passengers were accepted owing to lack of hotel accommodation within a convenient distance from the airport at Basle. In addition only temporary permits for overflights could be obtained from the Greek and Italian governments. In

45

July 1929 Major Brackley went to Cairo to replace Wolley Dod, the Station Superintendent, while the latter went on leave. As an example of the journeying necessary in those days his own experience deserves a mention Brackley flew to Paris in an Argosy and continued to Basle in an old Handley Page W10. From Basle he travelled to Genoa by train. A Calcutta flying-boat carried him to Alexandria via Naples, Corfu, Athens, Suda Bay and Tobruk. From Alexandria to Cairo he flew in a Hercules. He was in the air for 26 hours but the entire trip lasted five days.

In the autumn of 1929, when Imperial Airways refused to agree to the Italian demand for a pooled service between Genoa and Alexandria, Mussolini, the dictator of Italy, retaliated by banning overflights by the Company's aircraft. A new route through Vienna, Budapest, Skoplje, Salonika and Athens was tried. In the winter this choice of route was found to endanger the aircraft through icing, against which there was not at that time any effective safeguard. Consequently the Company was obliged to send their passengers from Paris to Brindisi by train. This journey of nearly 1,000 miles took two nights and a day. Another reappraisal of possible routes, which would avoid using Italian territory, induced Imperial Airways to shorten the distance between Athens and Alexandria by flying the Calcuttas from Athens to Mirabella (Crete), and onwards via Mersah Matruh on the Egyptian border.

To add to these troubles tragedy struck *City of Rome* when *en route* to Genoa, when engine trouble obliged the captain to put down at Spezia. A gale was blowing and the sea was rough but an SOS had been transmitted. An Italian tug found the Calcutta and took her in tow, but the towlines parted and in the darkness and bad weather the flying-boat capsized and sank with the loss of all on board. Four days later the other two Calcuttas were damaged after landing in rough seas off Mersah Matruh. After a few days they were back in service and early in 1930 two more flying-boats of this type were sent out from England to join them.

Looking ahead to the prospect of regular services to Delhi and beyond, Imperial Airways sent out a recently appointed Director, Lord Chetwynd, to take part in the negotiations to set up a subsidiary company that would include Indians on the Board. Chetwynd had been selected for this task because he was being considered as the future Chairman when Geddes retired. Unfortunately he caused great embarrassment by such tactless remarks as "Who would ever fly with an Indian?" On the flight home he refused to allow his pilot to wait for the mails at Baghdad, and two RAF machines had to be sent with them in pursuit. It was a costly blunder as revenue from the mails was bringing in £1,000 a week at the time. There were questions in Parliament and Geddes announced that Chetwynd was resigning as a Director on the grounds of ill health.

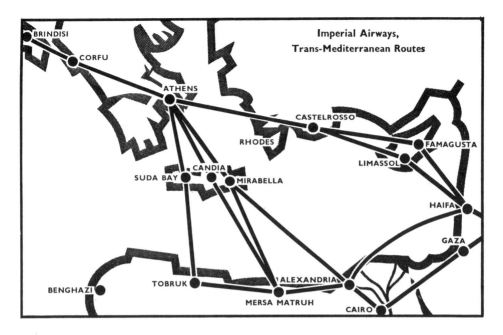

For quite different reasons Major Brackley also had his problems with the Board. When he arrived in Cairo in July 1929 he had expected to remain there only a few months. Eight weeks passed without any communication from London, not even when a Hercules crashed at Jask in Persia killing the pilot, a mechanic and a passenger. Brackley attributed the accident to misjudgement by a tired pilot working a very punishing schedule. In an editorial in *The Aeroplane* C. G. Grey placed the blame elsewhere:

> That fatal accident is another example of our muddled way of doing things. Whether a wingtip flare burst in the air and set fire to the wing, or whether in landing on flat featureless desert the pilot misjudged his height and crashed before the flare had gone out and so set the aeroplane alight, is irrelevant. The point is that the thing should never have been landing by its own lights. If there must be night flying it should be over a properly lighted route with proper flood-lights and beacons on the aerodrome. . .

The first word which Brackley received from Head Office was signed by Lieutenant Colonel Burchall, who had recently been appointed as Assistant General Manager, reporting to Woods Humphery. In a memorandum dated 16th October 1929 Brackley was informed that he would have to remain in Egypt, while Wolley Dod conducted a survey of the African route, commencing at Cape Town. He concluded: "I trust therefore that having stood the heat of the summer you will rather look forward to remaining in Cairo for

47

part of the season." Not surprisingly Brackley suspected that his prolonged absence from England had been intended all along, and he feared that his responsibilities and status were being diminished.

He was certainly not short of work. An Arab–Jewish war erupted in Palestine. Route deviations ordered by the British Government involved rescheduling of flights. On his own authority Brackley instructed pilots inbound from Bassa to avoid a night stop at Gaza, by carrying on to Cairo. After authorising night flying he was exasperated to receive a cable from Woods Humphery ordering him not to take the risk of doing so.

In December Brackley paid a short visit to London to see Sir Eric Geddes and to seek some reassurances about the extent of his authority as Air Superintendent. He was particularly incensed that his wife who had travelled to Genoa to catch the Calcutta flying-boat and join him in Egypt, had been told, incorrectly, that the aircraft was full. As a result she had been obliged to complete the journey by sea. Geddes had a reputation for impatience, particularly with human foibles, and Brackley had not mentioned the episode of his wife's journey but as he was leaving the Chairman's office Geddes remarked sharply: "I understand that your wife was with you in Heliopolis."

Although Brackley had been engaged in 1924 to act as the pilots'

Cabin of Short 'Kent' class flying boat which carried sixteen passengers across the Mediterranean at 105 mph. Three were built for Imperial Airways in 1931 and were known as 'Scipio' types.

spokesman this function had long since formed only a very minor part of the Air Superintendent's duties. He was principally engaged in route planning and development. Captain (later Air Commodore) G. J. Powell regarded his work in this sphere as "brilliantly done." In other respects, "Brackley was miscast, because he lacked the hard edge and short fuse that would have stopped some of our prima donnas and, believe me, we had plenty of them."

The Aeroplane noted Brackley's prolonged absence and the game of 'musical chairs' within his office in London:

> This business of Air Superintendent is baffling, Major Brackley was and still is really, but for many months Mr. Armstrong sternly occupied his official chair. Then one morning Mr. Wilcockson was there looking like Mussolini with a liver. A few days later the Air Superintendent was a positively portentous Mr. Walters…

Messrs. Armstrong, Wilcockson and Walters were all very senior pilots but Imperial Airways did not introduce the title of Captain until the early 1930s.

The ongoing problems of permits for a safe route over Europe to link up with the service to the east led Imperial Airways to try flying their passengers to Cologne. They continued their journey by rail to Athens where they boarded a Calcutta flying-boat for the flight across the Mediterranean. In December 1929 the service was extended from Karachi to Delhi using a de Havilland Hercules. Brackley had to remain in Egypt until April 1930, when the pilot W. Armstrong was sent there to be manager of the Near East Division. A year later the Company became aware that they were likely to face competition from the French Air Orient Company, which intended to fly to Saigon via Beirut, Baghdad, Karachi, Calcutta and Rangoon. Germany's response to the competition was to speed up the travel time by flying the sector from Baghdad to Basra at night.

In April 1931 Italy withdrew the ban on overflights and Imperial Airways was able to fly to Genoa, Rome and Naples. This concession was valid for a year after which a landplane service between Milan and Brindisi was planned. Once again the danger from icing was encountered in the winter conditions affecting northern Italy. There had to be another change of route and the cheaper option of sending passengers from Paris by rail was re-adopted. From Brindisi the flying-boats carried India bound passengers to Haifa, instead of to Alexandria, thereby shortening the route. This took effect from October 1931.

That year witnessed the introduction of the Short 'Kent' class flying-boats on the Mediterranean route, as a replacement for the Calcuttas. The vulnerability of the latter to rough conditions had shown the need for improved seaworthiness. With four Bristol Jupiter engines and an enclosed cockpit this was the first flying-boat with a stressed-skin metal hull to go into service. Known also as the 'Scipio' class, these flying-boats carried 15

Short Kent *Sylvanus* flown by Major Brackley from Medway, March 1931. In 1935 this aircraft caught fire during refuelling in Brindisi. *Robert Jackson.*

passengers at a cruising speed of 105 mph. By comparison with later years remarkably little testing was done before the Kents were delivered to Imperial Airways. Manufacture had only begun in October 1930 and in February 1931 the prototype made its first flight. *Sylvanus* was launched on 1st April and Brackley was flying it a week later.

The Kents displaced the Calcuttas which were transferred to service between Khartoum and Mwanza, but minor accidents delayed this plan. *Scipio* damaged a float in rough seas at Candia. She was back in service a few weeks later, but then her sister ship suffered damage to two engines when she was towed too close to another flying-boat in the harbour in Genoa. By August all three Kents were back in service. They earned great respect for their trouble-free performance, each one flying about 4,000 miles every seven days.

In 1932 passengers for India still had to travel from Paris to Brindisi by train. Thereafter they were flown to Athens, where there was a night stop, before flying on to Rhodes, Cyprus and the Sea of Galilee. In September a new service from London to Palestine and Iraq was introduced, to meet the steamships of Lloyd Triestino at Jaffa. To make the rail sector more tolerable Imperial Airways ensured that passengers travelled in first class sleepers, and they were allowed to break their journeys *en route* so that they could visit

50

places of interest in Europe as well as Cairo, Damascus and Jerusalem.

Pressure in Parliament for faster services grew as traffic across the Mediterranean increased, and criticism became more vocal when the mail from Palestine to the United Kingdom was transferred from the Company to KLM which completed the journey in two days' less time. Harold Balfour M.P. was a former RAF pilot and a director of both Spartan Airlines and Saunders-Roe, the aircraft manufacturer. He pointed out that the government's subsidy for the Cairo to Karachi route amounted to the equivalent of £180 per passenger, but the fare charged by Imperial Airways was £58. To the reply that this was true but irrelevant Balfour enquired about the grant to the Company of £20,000 for flying-boats. The House was told that the range of the Short Kents had been found to be insufficient and the grant was to meet the cost of extending it.

In the expectation of further obstacles being imposed by the Persians, Imperial Airways went ahead with their plans to move their route to the Arabian side of the Gulf. During 1932 this new route was opened to traffic. Much of the credit was due to Sheikh Issa who was on good terms with the RAF. He had persuaded most of the other sheikhs to allow flying-boats to alight along their coastlines and to co-operate in the construction of airfields for landplanes. The island of Bahrein became a staging post for both types of aircraft. Supplies for these could be brought in by ship and, in an emergency, vessels could be contacted by wireless.

In 1932 Imperial Airways seriously considered operating the whole route from Cairo to Karachi with flying-boats. Advice was sought from 203 Squadron of the RAF which was equipped with Short Rangoons and stationed at Basra. When it was explained to the Company how few suitable harbours existed the idea was put on ice. Instead the HP42s, were fitted with extra fuel tanks to improve their range. These flew from Basra across the delta of the Tigris and Euphrates rivers to Bahrein. The passengers were given dinner there before continuing to Sharjah where they stayed the night. Next day they flew over mountains before passing over the Gulf of Oman. To avoid infringing Persian regulations the route lay a few miles to the south of the coast. The last stop before Karachi was Gwadar in Baluchistan.

But as one problem was solved another had to be tackled. The government of India was reluctant to allow Imperial Airways to carry passengers and mail across their country *en route* to the Far East and to Australia.

CHAPTER FIVE

Cairo to the Cape 1919–34

After the conclusion of hostilities in 1918, and despite the multitude of demands upon its depleted financial resources, the British government was sufficiently enlightened to encourage and subsidise the clearance and levelling of a number of sites suitable for use by aircraft on the route southwards from Cairo. When the work was completed the Egyptian government accepted responsibility for those within its territory while the government of Sudan was paid an annual fee for the upkeep of their airfields. In 1920 the governments of the East African colonies were given custody of the aerodromes within their boundaries.

In the Union of South Africa, a company which in 1919 tried to organise an air service between Cape Town and several other cities was unsuccessful. In 1923 the South African government put all ten of the DH9s of its own air force on an experimental air mail service over a three month period to gain experience for a future commercial enterprise. Discouraged by the small amount of mail that was carried the government left it to private concerns to venture into civil aviation at their own risk. In 1926 the Junkers Company of Germany obtained from the Union government a grant for one year to provide Junkers aeroplanes and ground engineers. The pilots were to be recruited in South Africa. Financial problems of the parent company in Germany prevented this being realised.

In 1926 Royal Air Force squadrons based in Egypt conducted a number of navigation exercises to test their mobility and, in one of these, four Fairey aircraft flew to Cape Town, making twelve stops *en route*. This was so successful that on their return journey they were fitted with floats and continued their flight to England.

The first private concern in Britain to plan a commercial air service in Africa was one in which the Blackburn Aircraft Company had an interest. It was promised the co-operation of the governments of Sudan, Uganda and Kenya for an initial survey of the route to Kisumu, which lies on the shores of Lake Victoria. The colonial governors also provided financial assistance for an extension to Nairobi. Meanwhile the British government was giving financial support to the Burney airship scheme, which led to the construction of the R101. Blackburn's management pointed out to the Air Ministry that when the proposed route to India was finally opened by Imperial Airways and by airship, there would be a saving of eight days on the journey to East Africa if their own plans to develop services there were approved.

The Air Ministry was considering Blackburn's plans when in March 1926 Alan Cobham returned from a flight to Cape Town in a DH50J. His exploit was received with great admiration, but Cobham prudently predicted that, before a scheduled commercial service could become a reality, there would have to be airliners powered by three reliable engines. In addition one of the pilots on board would have to be a qualified navigator. The service would need to be backed up by a weather reporting system and the organisation on the ground would have to include facilities for rapid refuelling. The provision of small quantities of aviation fuel in a multitude of cans was not good enough.

The Air Ministry gave Blackburn a grant to conduct twelve round trip flights between Khartoum and Kisumu. Their pilot, Tony Gladstone, had completed two of these when his seaplane struck some flotsam as he was taking off from the Nile, and sank. At the time Sir Samuel Hoare and Sir Sefton Brancker were in Cairo on their way back to England from the inaugural flight to India. The Minister instructed the RAF to lend a Fairey seaplane to Gladstone, and Brancker travelled south to Khartoum to

Alan Cobham and Sir Sefton Brancker with their DH50 after their survey flight through Africa in 1926. *R. A. R. Wilson.*

accompany the pilot on a flight to Nairobi. All was going well, and the South African government had become interested in running a connecting service to Kisumu, when the Fairey seaplane was damaged once again by flotsam in the Nile. This delayed further trials until the autumn of 1926.

In November 1927 Alan Cobham flew out to Africa on a more extensive survey. He had been provided by the RAF with a Short Singapore flying-boat and had received financial backing from Sir Charles Wakefield. He ran into trouble landing in the harbour in Malta, where he lost a wingtip float, and the aircraft suffered further damage in the harbour in subsequent winter gales. When he was able to continue his journey he had fewer troubles and was back in England in June 1928. The following month Cobham merged his interests with Blackburn Airlines Ltd and submitted a scheme for a sched-uled service from Alexandria to Mwanza in Tanganyika, with an eventual extension to Cape Town. This had the support of the colonial governments.

Meanwhile Imperial Airways and the Air Ministry, represented by Wolley Dod and Lieutenant-Colonel Shelmerdine, were conducting a joint survey of the Cape to Cairo route working north from South Africa. In the House of Commons a Member pointed out that a more direct route from Britain to the Cape lay over the Sahara and the Cameroons. He was reminded that this passed over foreign territory. The Union government and the Company agreed that there should be a weekly flight in each direction from Cape Town and Cairo. Brancker had followed Blackburn's efforts in Africa with interest and encouragement, but the British govern-ment was under an obligation to give first refusal to Imperial Airways on any new route.

The clash of interests was resolved by Sir Samuel Hoare's proposal that Imperial Airways should absorb Blackburn–Cobham for a sum of £50,000. The Company's disbursement was in turn balanced by a new subsidy for its African services. The Union government was also ready to subsidise the service for five years, and all the Colonial governments promised a financial contribution. The Beit Trust was also generous. Sir Alfred Beit had been an associate of Cecil Rhodes and had established a trust fund to develop communications in Africa. The Trust adopted the suggestion of the pioneer airman Campbell Black, that emergency airstrips could be made far more cheaply by widening existing roads than by building individual ones. In an atmosphere of co-operation and goodwill Imperial Airways (Africa) Ltd was registered as a private company in June 1929. Nevertheless a very consider-able amount of work had to be put in hand before flights to Cape Town could begin.

The route south from Cairo passed through territory which was under the Crown or within British political control. This reduced the political obstruction that was proving such a problem elsewhere but the distance

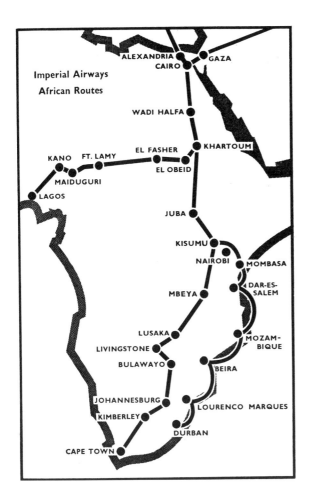

Imperial Airways
African Routes

involved was not far short of 6,000 miles. Given the short range and slow speed of the aircraft then in service almost sixty aerodromes or landing grounds needed to be prepared. During the rainy season many of these would be rendered unsafe for landing and take-off. The earth surface between Khartoum and the Sudan–Kenya border was often a swamp. A chain of wireless stations had to be put in place so that aircraft could remain in contact. A meteorological service had to be organised.

Imperial Airways had to build rest houses for passengers, flight crews and their local staff. The Company had also to train about twenty-five airport managers to take charge of the new staging posts. Some of the aerodromes were situated thousands of feet above sea level which, combined with very high daytime temperatures, made it necessary to provide an extra long area for aircraft to build up sufficient speed to become airborne. This is as true today as it was then. In 1930 Imperial Airways ordered eight four-engined aircraft from Armstrong Whitworth, designed specifically for the African route. These had to be able to cruise at 9,000 feet with one of their engines shut down.

Known as Atalantas they were the first monoplane airliners to be built in Britain. They were still under development when in 1931 the first service southbound left Cairo, on a route network which extended only as far as Mwanza.

The perennial problem of a shortage of aircraft restricted Imperial Airways to the use of two Argosies on the sector from Alexandria to Khartoum. Only two Calcutta flying-boats were available to carry the passengers and mail on to Mwanza. Inevitably sandstorms, rainstorms, unserviceability, poor ground communications and delays awaiting spare parts, caused such disruption to schedules that the colonial authorities threatened to discontinue subsidies to the Company. Assistant General Manager Burchall was sent out from England to investigate and to try to explain to the impatient East Africans why it had been so difficult to keep to the published timetable. Delays to the mail had upset them more than complaints from passengers. The latter were more understanding and sportingly saw themselves as air pioneers, sometimes obliged to sleep under canvas at unscheduled stops and, in flight, to accept sandwiches rather than normal meals because too few restaurants existed for the Company to be able to supply anything else.

On 20th January 1932 an HP42 left Croydon to initiate the new air mail service to Cape Town. In sequence an Argosy, a Kent and a DH66 Hercules were used to carry the mail to Cairo. A DH66 of mid 1920s vintage completed the journey to the Cape via M'beya, Salisbury and Johannesburg. The whole 8,000 mile journey took ten and a half days. But as had happened before bad weather prevented adherence to the schedule and the Hercules arrived 18 hours late at Nairobi. Tropical rainstorms caused further delays and the mail reached Cape Town two days late. The inaugural northbound Hercules taxied into a badly filled hole at Salisbury and was damaged. A relief aircraft which arrived from Cairo made a forced landing and nosed over in soft ground. The mail finally arrived nine days late in England. *The Aeroplane* was philosophical, declaring that eight new Atalantas would shortly enter service "and similar comedies should be few and far between."

The second southbound aircraft was forced to make an unscheduled landing in foul weather and two days passed before searching Puss Moths found her. The airline crews had every reason to fear tropical storms. Tony Gladstone and his co-pilot, Kidston, had recently been killed when their aeroplane had broken up in clear air turbulence over Natal. Kidston had

Chairman Sir Eric Geddes (left)
takes leave of Sir George Beharrel
before his flight to South Africa
in December, 1932. Major Woods
Humphery (centre) looks on.

earlier flown a Lockheed Vega from England to the Cape. In April 1932
passenger bookings were accepted both in England and South Africa, but as
the sectors between Cairo and East Africa were always fully booked, no
through bookings could be allowed. When the service was properly established the Cairo–Cape Town sector was flow by HP42s and Atalantas until the
advent of the Empire flying-boats in 1937.

The Atalantas had still not entered service at the end of 1932 when Sir
Eric Geddes and Major Woods Humphery travelled together to the Cape to
gain first hand experience of the route. They flew out in an HP42 Hannibal
to show off this aircraft, which was flown by G. J. Powell, subsequently an Air
Commodore in the RAF and, after the Second World War, operator of his
own air line. Another reason for their journey was to discuss with the
statesman, General Smuts, what effects the world financial crisis was having
on the South African economy, and the likely response of the Nationalist
party government to the Company's plans to carry passengers and mail to
and from the Union.

An important result from their trip was the formulation of the Empire Air Mail Scheme. On their way south, Woods Humphery explained to Geddes a plan proposed by the Company Secretary, S A. Dismore, for Imperial Airways to carry all letter mail for the Empire without an air surcharge, but with the Company being paid an economic rate by the British government. The first hurdles would be to convince the General Post Office and the Air Ministry. Woods Humphery impressed Geddes with his argument that the current revenue from surcharged mail was quite insufficient to permit the airline to expand as it should. The Chairman was convinced and the two men decided to return to England by sea, to allow themselves uninterrupted time to work out the details of a memorandum to be submitted to the cabinet. As is usual in a parliamentary democracy two years were to pass before their initiative produced results.

In January 1933 Major Brackley set off on a proving flight to the Cape with the first of the Atalantas to be delivered. This was the first airliner in Imperial Airways to have a cruising speed in excess of 100 mph. But with space for only ten passengers, at a time when there were local demands that the number of flights should be doubled, it is clear that they were not large

The civil air ensign flies from a Handley Page 42. *Croydon Airport Society.*

enough for the African service. Consequently the Handley Page Hannibal was put on the route, but only after the airfields at Juba and Malakal in the Sudan had been strengthened by the Shell Oil people, who had discovered how to build roads over the swampy ground. Imperial Airways paid them £8,000 for each of the airfields.

This was an era when the passengers were roused around dawn for a long day's flying that concluded at dusk. The Company did their best to provide a good night's rest. A popular night stop was Luxor as time was allowed for a short visit to the Valley of the Kings, and the temples of Luxor and Karnak. A rest house was also begun at Pietersburg in Transvaal.

In the face of continuous criticism that private pilots had been flying to and from Africa in half the time taken by the Company, Imperial Airways attempted to improve upon their schedule by eliminating night stops and changing refuelling places. But the principal obstacle to a speedier time-table was the time-wasting railway journey from Basle to Brindisi, resulting from the French and Italian prohibition on commercial overflights. In February 1934 the Italian government finally reached an agreement with the British government on a range of issues which permitted Imperial Airways to fly across Italy, and by early June the railway journey from Switzerland to Brindisi could be dropped.

In Kenya Wilson Airways, operating out of Nairobi, was useful to Imperial Airways as a feeder service. So too was Rhodesia and Nyasaland Airways (RANA) which began operations in 1933 and brought traffic from Beira in Portuguese East Africa. Union Airways in South Africa had been founded by English residents and this airline, and another which was a subsidiary of Junkers, also fed passengers and mail to Imperial Airways. Then Boer nationalism and anti-British sentiment was brought into play. The South African government wanted the mail to be carried by a company run by South Africans. The Junkers Company underbid Union Airways which sold out to it. The immediate consequence was that Imperial Airways was allowed to carry only Empire mail, and national mail was reserved for the local carrier.

In January 1934 the Union government took further measures to assert control over airline activity within its borders. It took over Union Airways, making a token payment to its creditors among whom the Junkers Company predominated. In April the South Africans refused to renew the mail contract with Imperial Airways, denying passage to the Company's land-planes beyond Salisbury in Southern Rhodesia. In a crafty move designed to put RANA out of business the Union government invited Imperial Airways to operate a flying-boat service from East Africa to Durban and Cape Town. Realising that this new service would be taken over when it became profitable Imperial Airways did not accept this offer.

During 1934 moves were afoot to bring into being the Empire Air Mail Scheme (EAMS). In May Sir Kingsley Wood, the Postmaster-General, appeared at Croydon airport to present to Imperial Airways the new royal air mail pennant. It made its first journey on the HP42 *Hengist* to Paris carrying the Indian mail on the first stage of its journey. Blue pillar boxes specifically for air mail letters next made their appearance on the London streets and their contents were rushed to Croydon in a blue van. Letters to India cost sixpence per half ounce. Those carried to South America on the French or German airlines cost nine times more than this. When putting their case to the government in favour of the EAMS, Imperial Airways had declared their interest in the future operation of flying-boats, whenever possible, on the Empire routes. One advantage to be gained was the avoidance of the expense of enlarging the existing aerodromes. The EAMS was approved by the cabinet and by the end of the year was accepted by Parliament. The Air Ministry agreed to make marine facilities available.

This was the signal for Imperial Airways to order from Short 28 four-engined flying-boats. Major Mayo had drawn up the technical specifications and invited tenders from several manufacturers, but only Short had responded. These aircraft were ordered off the drawing board and would go into service with no pre-testing of prototypes. Costing £61,000 each they were designed to cruise at 160 mph with a range of 760 miles. This was by far the largest single order ever placed by the Company. In the past their habit of ordering aircraft in very small numbers had put up their cost because every one had to be hand-built.

The satisfactory conclusion of an agreement with the Italian government about overflights turned the attention of Members of Parliament, and in due course the British government, upon the French. Questions were asked about the right of French aircraft to continue to fly unhindered over British territory in the face of the procrastination of their government over this issue. Sir Philip Sassoon, Under-Secretary of State for Air, and Lieutenant-Colonel Shelmerdine visited Paris to press for a route across France for British commercial aircraft. The implied threats in the House of Commons must have had their effect because the ban on overflights was withdrawn. Sadly a shortage of aircraft prevented Imperial Airways from introducing a passenger service without delay, but a mail service across France and Italy to Brindisi by DH86 was opened, and facilities for a flying-boat base at Marseille were approved by the end of 1934.

In Southern Africa also progress was made. The government of Portugal agreed to allow Imperial Airways the use of ports in their East African colony for their flying-boat service, and in June 1937 the Empire flying-boat *Canopus* reached Durban with passengers and mail.

CHAPTER SIX

The Growth of Domestic Competition

The Wall Street Crash of 1929 caused a severe and long-lasting depression world-wide, not least in Europe. There was a big drop in the number of American visitors, and load figures on all air routes decreased. For some time the European service offered to the general public by Imperial Airways remained in a sorry state of stagnation. There were still regular flights to Paris, Brussels, Zurich and Cologne, just as there had been before the original airlines had amalgamated, but no others. These services were not profitable and could not compete in the matter of speed. Nor could the Air Ministry provide additional funds to allow the Company to offer a serious challenge to the heavily subsidised foreign airlines. It remained government policy that services to the Empire should have the first priority, and that no taxpayers' money should be allocated to any other British airline that might wish to operate on routes neglected by the 'chosen instrument'

Keith Granville, who was to become Chairman of the British Overseas Airways Corporation (the successor to Imperial Airways) in 1971, had joined as a trainee in 1929 at a salary of ten shillings a week. Head Office was then in Charles Street between the Haymarket and Lower Regent Street. The few staff who worked there included two baggage porters. The commercial hub of the business was in the basement which housed a reservations office with half a dozen telephones, including one with a direct line to Thomas Cook. The table in the small boardroom was of such modest proportions that Granville eventually used it himself as a desk during his last ten years of service. He recalled the difficulties the Company sometimes had in meeting the weekly wage bill. Cash to pay the staff at Croydon was obtained by cashing a cheque at a West End branch on Friday, in the knowledge that it would not be cleared through the bank at Croydon until Monday. Everything depended on revenue from the week-end flights to Paris. Phil May who had joined in 1928 remembered that there were only about thirty staff in the entire accounting department, where they sat on stools and pored over ledgers: "If so much as two shillings was missing there was an inquest." At Croydon the total fleet of 21 aircraft was valued at under £500,000.

When Major Brackley returned to England in April 1930 his suspicions that his extended stay in Cairo had been devised to alter his terms of employment were confirmed. The Assistant General Manager, Colonel Burchall, offered him a post in South Africa. A few days later, after he was granted an interview with Woods Humphery, the latter proposed his appointment as Manager, European Services. But Brackley was determined to remain Air Superintendent with direct access to the Board.

In May, in a letter to Woods Humphery, he put his case:

> "Since the inception of the Company over six years ago I have received no advance in salary or ex-gratia payment of any kind. Quite apart from my ordinary work I think you will be the first to admit that I have been the sole means of obtaining for the Company a number of remunerative special charters and a large number of passengers (often at great personal inconvenience) in addition to the very profitable new German deal put through early last year. As a result of the latter the Company is now drawing handsome fees as consultants. I feel sure that if these facts are placed before the Board, with your kind recommendation, I shall obtain some financial recognition of my past work and that the Board will not regret such actions."

Brackley failed to receive any such financial recognition but within a week he was restored to his post as Air Superintendent.

Sir Eric Geddes was able to report a profit of just over £60,000 for the year 1929/30, but was unwilling to give details of the effects of the subsidies which the Company received, because of the incorrect deductions which would unfairly be made, "We have the paradox of comparing subsidy with profit," he said. "The Air Ministry increases our subsidy on condition that we put by this tremendous obsolescence [depreciation] of 20 per cent and so increase our costs. The Inland Revenue declines to allow the obsolescence in full as a legitimate cost, but taxes us on it as they say it ought to be a profit. One government department says it's a cost and another says it's a profit."

Imperial Airways suffered another accident in October 1930 when a seven year old twin-engined Handley Page W8 flew into the ground in poor weather near Neufchatel. There were fatalities, but the event was over-shadowed by the greater disaster to the airship R101 carrying the Secretary of State for Air, Lord Thomson and Sir Sefton Brancker to an Imperial Conference in India. Only eight of the 54 people on board survived.

Passenger load had started improving in 1931 when eight Handley Page HP42s, half of them destined for European services, were delivered. These were the 'Heracles' class, seating 38 passengers. The 'Hannibal' class carried 24 on the eastern route allowing extra fuel tanks to be installed. They were the first four-engined airliners to be built in Britain, and the first to fly in regular service anywhere in the world. They were also the first Company aircraft to have an enclosed cockpit since Captain Hope had wielded his axe.

The pilots christened the HP42 "The Flying Banana", and there was so much spring in the fuselage that they needed to keep their harnesses securely fastened when they taxied over uneven ground. The passengers found that they were quiet and comfortable, not least because they sat some distance from the noise of the engines, and two stewards were on hand to

The airship R101 in 1929.

serve them. Refreshments had to be paid for and the pilots were charged for their cups of tea or coffee also. *The Aeroplane* stated that "there were no bumps when airborne. We doubt that there are more comfortable machines in the world than the HP42." Certainly the 'Silver Wing' service to Paris offered a superb cuisine but the aircraft was very slow. A cruising speed of 95 mph was far inferior to what was being achieved in the United States where Boeing were offering American airline operators a monoplane with a retractable undercarriage and a cruising speed of 150 mph. Within a few years the Douglas DC3, still flying in many parts of the world today, would appear in the skies.

In the summer of 1931 Dismore was flying the HP42 'Hannibal' when a propeller disintegrated, fragments striking the propeller above it, causing severe vibration and loss of power. He was flying near Tonbridge and tried

IMPERIAL AIRWAYS

DAILY DEPARTURES AND ARRIVALS AT THE AIR PORT OF LONDON (CROYDON)
of the services operated by Imperial Airways, the German Air Lines, and the Belgian Air Lines

OUTWARD BOUND Departure from Croydon	SERVICE 15 OCTOBER to 31 OCTOBER, 1931	INWARD BOUND (Corresponding Service) Arrival at Croydon
a. 08.30 hrs Weekdays only	Paris by *Imperial Airways*	10.45 hrs
b. 09.00 hrs Weekdays only	Amsterdam, Hanover, Berlin. This service is operated by the German Air Lines (Deutsche Luft Hansa)	17.10 hrs
c. 09.30 hrs Weekdays only	Brussels, Cologne, with connexions for Antwerp, Dusseldorf and Essen. This service is operated on alternate days by *Imperial Airways* and the Belgian Air Lines (S.A.B.E.N.A.)	14.25 hrs
d. 12.30 hrs Daily including Sundays	The 'Silver Wing' Service to Paris by *Imperial Airways*	14.45 hrs
e. 12.30 hrs Wednesdays only	THE AFRICAN AIR MAIL France, Italy, Greece, Egypt, Sudan, Uganda, Kenya and Tanganyika by *Imperial Airways*	10.45 hrs Fridays
f. 12.30 hrs Saturdays only	THE INDIAN AIR MAIL France, Italy, Greece, Palestine, 'Iraq, Persia and India by *Imperial Airways*	10.45 hrs Tuesdays

TELEPHONE AIRWAY TERMINUS—VICTORIA 2211 *(Open Day and Night)*

for information and reservations on the services operated by Imperial Airways, the German Air Lines and the Belgian Air Lines. Connecting cars for these services leave *Airway Terminus* 45 minutes before the departure time from the Air Port of London (Croydon), viz., (a) 07.45 hrs, (b) 08.15 hrs, (c) 08.45 hrs, (d) 11.45 hrs, (e) 11.45 hrs, (f) 11.45 hrs., but passengers are asked to arrive at *Airway Terminus* 10 minutes ahead of these times, in order to allow of the necessary examination of tickets, passports, and the registration of their baggage

N.B.—*Airway Terminus* is situated under the canopy facing the entrance to the Continental Departure Platform at Victoria Station, London, SW1

This Time Table is based on the 24-hour clock—that is to say, the terms a.m. and p.m. are not used, but the hours, starting from midnight, are numbered consecutively from 0 to 24, covering the period from midnight until midnight. 8.0 a.m. thus becomes 08.00 hours, while 8.0 p.m. is written as 20.00 hours, the last two figures in the group indicating minutes past the hour—for example, 8.15 a.m. is 08.15 hours; similarly, 8.25 p.m. becomes 20.25 hours. In short the times are always expressed in four figures, the first pair representing hours and the second pair minutes.

This card is printed in England and published normally every month by Imperial Airways at *Airway Terminus* and is issued for the convenience of travel agents.

IA/a/37h 2,000 10/31

to avoid houses in his attempt to land in a field. In so doing he collided with a telephone pole and a tree stump. Everyone walked unhurt from the aircraft, astonished to observe how severely damaged it was. Handley Page had considered the possibility of propeller fracture, and had located the mail and luggage compartments in line with the blades.

Radio telephony between control tower and aircraft lay in the future. To the disdain of the pilots, the airport authorities at Croydon instituted a system whereby a departing captain signalled to a man standing alongside that he was ready to taxi. This individual was required to wave a green flag at the control tower and the controller displayed a visual signal to the pilot when it was safe for him to taxi away. This system was abandoned when the pilot W. Rogers put on a porter's peak cap and walked out to the tarmac with both red and green flags, blowing a whistle.

Late in December 1931 when Imperial Airways held its seventh annual general meeting Sir Eric Geddes disclosed a substantial drop in profits to £27,140. Shareholders were comforted by the statement that aircraft numbers had risen to 41 including flying-boats. This total had however been inflated by the inclusion of three HP 42s and eight Armstrong Whitworth Atalantas still under construction as well as three old twin-engined Handley Page W8s and two charter machines. In Britain Imperial Airways had a reputation for safety and reliability. Attention to passengers' requirements was good but there was continuing criticism of the slow speeds of their aircraft in comparison with KLM and other European airlines. Part of the trouble was the tiny number of aircraft ordered at any one time by the Company. At £42,000 each even eight HP42s did not set the imagination of the manufacturers alight and Imperial Airways was obliged to buy only British equipment. In 1932 in response to an article in the *Daily Mail* Sir Eric Geddes provided the following information:

	aircraft employed in air transport	average payload (lbs)	pilots employed	weekly air mileage
France	269	1,380	135	109,000
Germany	177	1,849	160	299,000
Italy	77	1,836	61	60,000
Great Britain	32	4,858	32	38,000

At Croydon the new terminal had been completed, but except for those staff who lived alongside the airport or possessed a bicycle, there was still a fifteen minute walk from Waddon Station as the bus service was very infrequent. Nor had much been done to enable pilots to arrive or depart from the aerodrome in conditions of poor visibility. It was the idea of one of Imperial Airways' original pilots, Dismore, to have a long white line painted in chalk across the grass for pilots to follow when taking off in fog. W. Armstrong recalled this in *Pioneer Pilot*:

> Sometimes the fog was so thick that we had to be led out to the beginning of this line by a tractor. We would give a signal to the tractor driver to start and we would slowly move off from the tarmac apron while we took care to keep him in sight. We would turn onto the white line, moving gently to be sure that we were running true and would not lose the line at speed, then gradually open up our engines and take off along the line. This technique became quite a regular practice during the winter fog. With visibility as low as ten yards we would still use this white line with assurance.

Not surprisingly passengers were often apprehensive about landing at Croydon when the word 'fog' was mentioned. A woman who asked the pilot W. Rogers how he would be able to find his way was told that it was his invariable practice to stick his head out of the cockpit and take a good sniff, and when he could smell the Beddington sewage works he knew he was very close to the aerodrome. The humour of this former non-commissioned RFC pilot, known to his colleagues as 'Cockney Rogers', was not always appreciated by the passengers, nor by the management and was to affect his career.

Even without any prospect of a subsidy from the Government there were individuals and organisations preparing to offer air services in the British Isles and beyond, and to take advantage of the opportunities for profit neglected by Imperial Airways. In 1931 The Whitehall Trust Ltd, which had been founded by Lord Cowdray and managed by the Hon Clive Pearson, acquired shares in the aircraft manufacturers Saunders-Roe of Cowes. The Whitehall Trust already had shares in another manufacturer, The Spartan Aircraft Company, and this interest in aviation was probably at the instigation of one of their members, Captain Harold Balfour M.P., who had served with distinction as a pilot during the war.

The four main railway companies, the Great Western (GWR), the London Midland and Scottish (LMS), the London and North Eastern (LNER) and the Southern (SR), became seriously concerned at the competition being encountered from the increasing number of long distance coach operators. They foresaw a time when domestic airlines would also make inroads into their profitability. For this reason they had obtained from Parliament permission to operate air services within their own territories. The Southern Railway, already affected by the competition from the airlines on their cross-channel services, went so far as to acquire a substantial number of shares in Imperial Airways.

In 1933 the Great Western Railway came to an agreement by which Imperial Airways provided a Westland Wessex three-engined monoplane together with a flight crew and engineers. This aircraft which had seats for six passengers made its inaugural flight between Cardiff and Plymouth, the pilot being Gordon Olley. The Postmaster-General authorised the carriage

Passengers study airport and route weather repots in Croydon's waiting room. 1933. *Croydon Airport Society.*

of letters by air at a surcharge of threepence each. Heavy luggage could not be fitted in the aircraft but it was forwarded by rail and delivered to its destination for no extra charge.

This was a GWR enterprise, but the following year all four railway companies came to an agreement with Imperial Airways by which the latter would supply the flight crews and technical support. Sir Eric Geddes justified this to the shareholders as a sensible arrangement, given the fact that the railway companies had been granted powers to operate to the Continent, whilst facing competition from airlines within the British Isles. As a result Railway Air Services came into being and de Havilland obtained successive orders for the twin-engined DH84 Dragon and DH89 Rapide, and the four-engined DH86 which carried ten passengers.

A further agreement between the railway companies and Imperial Airways resulted in the former warning travel agencies, most of whose business was in rail tickets, that the railway companies would no longer recognise them if they dealt in airline tickets other than those of Imperial Airways and the foreign airlines. In effect from 1933 this proved to be a serious bone of contention to the new domestic airlines and in the following years 21 questions were put in Parliament. Complaints were also heard in speeches in the debates on the Air Estimates. At the end of 1933 Geddes could report that traffic had risen by over 75 per cent and that Imperial Airways had flown over two million miles, almost all on regular service. The profit for the year had risen to £52,894.

As the winter set in the Air Ministry introduced a system of air traffic control at Croydon, restricting entry within a ten mile radius to aircraft equipped with wireless, if visibility was under 1,000 yards horizontally or 1,000 feet vertically. Outside the ten mile zone pilots continued to fly in bad weather and in darkness, trusting to luck that they would not collide with any other aircraft. On 30th December an Avro 10 flying from Cologne to Croydon hit the 900 foot wireless mast near Bruges, and dived into the ground killing all on board. The Avro was nine miles off course and flying too low. The relatives of a passenger claimed £40,000 damages, attributing negligence to the pilot.

Two landplanes, *Scylla* and *Syrinx* entered service in 1934. These had been ordered from Short's, because Handley Page had closed down their HP42 production line and quoted too high a price for Imperial Airways to order two more. Major Mayo, the technical manager, had therefore decided

Scylla was one of two landplanes which entered service in 1934. They were built by Short's using Kent class flying boat wings on a new fuselage. Still in service at the outbreak of war, 1939. *John Stroud.*

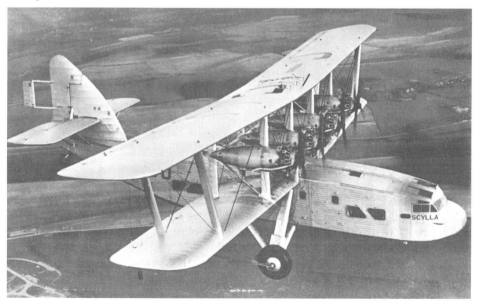

upon the fitting of Short Kent flying-boat wings on a new fuselage. These two airliners, the last biplanes built by Short's, had the same Bristol Jupiter engines as the HP42 and were very strongly built, but they were detested by the pilots for their heavy handling qualities and unpopular with the mechanics who had to service them. Carrying 34 passengers and cruising at under 100 mph they did not enjoy the approval of passengers, who complained that they wallowed around like a small boat in a rough sea. *Scylla* was damaged after a landing at le Bourget when the wheel brakes jammed. It skidded along the grass on its nose, the tail approaching a vertical position. The station engineer, Lloyd Ifould, described the outcome: "We found the steward and most of the passengers in a heap in the cockpit, mingled with the contents of a well-stocked bar which followed them through the machine in their headlong flight to the cockpit."

With no encouragement whatever from the government other airlines were being formed in the United Kingdom. Gordon Olley left Railway Air Services to found Olley Air Services, while the remarkable man who founded Hillman Airways had little formal education and had been a farmer's boy before he joined the Army. On demobilisation in 1918 he had reached the rank of Sergeant-Major. Thereafter Edward Hillman was in turn a chauffeur in the diplomatic service, a taxi driver, and the operator of a car hire service. In 1928 he bought a motor coach and later the same year started Hillman's Saloon Coaches in Essex. By 1931 he claimed to possess 300 coaches which toured seaside towns.

Forced out of business by government curbs on the proliferating bus companies Hillman used the £145,000 compensation which he received to found his own airline. By December 1931 he was operating a de Havilland Puss Moth from a small grass field near Romford which he rented from a local farmer. By the following April he had added two DH83 Fox Moths which made regular flights to Clacton-on-Sea. Determined to keep his costs as low as possible he hired former RAF pilots, who had not held commissioned rank but loved flying, and paid them little more than he had paid his bus drivers. In fact it was said that they looked like bus drivers, and Hillman did not regard them as any different. "I don't want no high-falutin pilots and toffee-nosed flying hostesses," he said, "I am going to run to Paris like a bus service from my field at Maylands, you'll see."

A 'workaholic' who was seldom seen in a respectable suit he cursed his pilots when they told him that the weather was too bad for flying, but was on hand with a mug of steaming tea when a pilot landed after many hours at the controls. "Here you are, mate," he would say, "I reckon you deserve this." It was less a mark of gratitude, one pilot recalled, than a device to keep his pilot on the job while some further passengers were hurried out to the machine." Hillman bought his first DH84 Dragon off the production line and used to claim that it had been designed to his specifications. He did not want an aircraft which required more than one pilot and scrutinised his bill of almost £3,000 with great care. When he noticed the charge of £50 for the toilet facility he exploded in anger: "Fifty quid for a bleeding shithouse!"

By 1933 his aircraft were flying to Paris at a lower fare than Imperial Airways, and they completed the journey in thirty minutes' less time. Within two years the frequency had reached four daily flights with others to Brussels and le Zoute. There were no frills. Meals were not served in the air and the passengers flew from Romford, but this was no further from Central London than Heathrow is, and road traffic at that time was much lighter. Hillman also operated services to Liverpool, Belfast and Glasgow and had a contract from the GPO to carry mail.

On 19th December 1934 Hillman Airways became a public company and all the shares were taken up within an hour. Twelve days later Hillman died from a coronary attack. His executors had to sell his own large block of shares on behalf of the family. These were purchased by the banking house of Erlangers one of whose Directors, Gerard d'Erlanger, was later to be the Chairman of the British Overseas Airways Corporation. Major R. McCrindle was appointed Chairman of Hillman Airways.

In 1934 Spartan Air Lines began operations in association with Southern Railways between Croydon and Cowes. Their inaugural service was flown by P. Lynch-Blosse in a Spartan Cruiser. Another extremely efficient company was British Continental Airways which was owned by the Chairman of Lloyds, Sir Percy MacKinnon. It began services to Ostend, le Zoute and Brussels in 1935 and pioneered the route to Malmö, but suffered from the bookings ban enforced by the railways on travel agents.

Imperial Airways had in 1934 flown more than seven and a half million miles which was well beyond the distance required to qualify for its subsidy. At the tenth annual general meeting Sir Eric Geddes reacted to criticism of the Company's slow schedules, compared with the speed achieved by the successful competitors in the recent Mildenhall to Melbourne Air Race:

> All the machines in the Race flew free from the hampering regulations of Customs and Passports which would be applied in the ordinary way to passengers and freight ... the Douglas DC2 carried twice the amount of fuel she would normally carry. That saves stops. It flew by night over a route mainly unlighted and no commercial company would care to do that in regular passenger service, nor does the Douglas provide comfort for long flying as given by our Hercules.

In April 1935 the Company and Swissair jointly operated services between Croydon and Basle, the British airline making a call at Paris *en route*, the Swiss putting into Lille. The latter's timetable was a witness to the vastly greater speed of their airliners as compared with those of Imperial Airways. Another service was inaugurated to Budapest via Halle, Leipzig, Prague and

This KLM DC2, second in the air race to Melbourne in 1934, revealed the supremacy of United States aviation technology and the obsolescence of Imperial Airways airliners.

Vienna. This was run in conjunction with the Belgian airline, Sabena. The Company also obtained permission to fly across Italy to Brindisi to link up with its Empire services.

By 1935 most of the senior captains of Imperial Airways were veterans who had flown during the war. The younger co-pilots, ranking as first officers, had obtained their flying experience before entry to the Company, either in the RAF five year short service commission scheme, or in the Reserve of Air Force Officers. A few had been sufficiently wealthy to pay for their own training. Imperial Airways required pilots to possess a licence in wireless telephony in addition to a 'B' (Commercial) pilot's licence, and a navigator's licence. In case it was necessary to sign out an aircraft after maintenance at an airport where no Company engineer was stationed, a ground engineer's licence was a further requirement.

The pilots liked to meet in the buffet in the main reception hall of the Air Terminal at Croydon, to drink when their working day had been completed. Visitors also liked to go there, some in expectation that they might recognise airmen whose adventurous lifestyle had been reported in the popular newspapers. Woods Humphery was uneasy about the impression that their cheerful carousing might give to nervous passengers. Brackley gave an order to the Company's pilots that the buffet was out of bounds. This was greatly resented but the Aerodrome Hotel obligingly opened a pilots' bar where they could refresh their thirst in privacy. They formed a committee to run it as a club with an annual subscription. In time the walls of the pilots' bar were hung with caricatures of a number of their members such as O.P. Jones, Herbert Horsey and 'Scruffy' Robinson. The latter was also known as 'the undertaker' because he had once crashed in a cemetery.

The gossip columns of some newspapers began to include the names of celebrities who flew to and from Paris, usually those familiar to the public from appearances on stage and screen, along with boxers, jockeys and racing drivers. Sometimes a travelling companion of one of these was named. This publicity may have been more welcome to Imperial Airways than to some of the individuals identified, whose journey had been made for private reasons.

In the summer of 1935 Sir Eric Geddes was made aware that the monopoly enjoyed by Imperial Airways was in jeopardy. Lord Londonderry was replaced as Secretary of State for Air by Viscount Swinton. A number of interested parties began to exert pressure upon the government to relax the policy of supporting only one 'chosen instrument'. In August 1935 Viscount Swinton visited Sir Eric Geddes and suggested that Imperial Airways should surrender its exclusive right to operate north of the London–Cologne–Budapest line. Furthermore the government should be free to offer financial support to another air company in that area without compensating Imperial Airways.

The Minister was most concerned to emphasize how faithfully, efficiently and generously the Company had performed its duties, and that it fully deserved the confidence and support of the government and Air Ministry. However, although he claimed not to know the reason, he warned Geddes that there was dissatisfaction both within the government and the House of Commons over the Imperial Airways monopoly. Viscount Swinton added that this dissatisfaction amounted to hostility, both among Members of the House of Commons and outside. He concluded by suggesting that this hostility should be countered by permitting some other operator to enter the area reserved to Imperial Airways in the existing agreement.

To Geddes' rejoinder that these proposals conflicted with the principles of the Hambling Report, Swinton replied that no Report remained irrevocable policy for all time and he indignantly denied Geddes further conjecture that the Company was the victim of changes at the Air Ministry, and was drifting from being the chosen civil aviation agent to becoming one of a group of competing airlines. Nevertheless a government committee headed by Sir Warren Fisher recommended that a second British airline should indeed operate in Europe and be given a subsidy to do so.

In February 1936 Imperial Airways had no option but to give up its claim to the area north of the line London–Berlin and the government signed a mail contract with Whitehall Securities (as the former Whitehall Trust had become) and gave them a subsidy. But whereas Imperial Airways had always been under an obligation to operate aircraft manufactured in Britain the newcomer was given the freedom to purchase abroad if no British aircraft capable of cruising at 200 mph was available. Re-armament against the threat of Hitler's Germany had already begun and aircraft manufacturers were fully engaged with contracts to supply the RAF. Consequently this airline bought Fokkers for their services to France, chose the Junkers Ju52 for the route to Hanover and ordered Lockheed Electras from the United States for their services to Paris and Malmö.

Thereafter Whitehall Securities swallowed up Hillman Airways, Spartan Air Lines, British Continental Airways, Highland Airways and Northern and Scottish Airways, to emerge as the first British Airways Ltd. The Managing Director was Ronald McCrindle and the other Directors included Gerard d'Erlanger and Harold Balfour M. P. The shares were all held by private interests, the government having no financial stake whatever. The new company invited RAF pilots with over 1,000 hours flying time to apply, offering them jobs as second pilots whilst they studied to obtain navigators' licences. A team was sent to survey the routes from London to Amsterdam, Hamburg, Copenhagen and Malmö. At the time there was no British airline operating a service east of Cologne, whilst the Scandinavian countries were able to send all first class mail by air.

Crilly Airways, a Midlands firm, had plans to operate to Lisbon with an eventual extension to Gibraltar and West Africa. Prime Minister Stanley Baldwin thought highly enough of this venture to send a letter by the airline

Fokker F.12 operated by British Airways, 1936, one of several types of airliners which outperformed those of Imperial Airways on European routes.

FLY TO LE TOUQUET IN LUXURY
for the day or for the week-end
(Roulette is now played there)

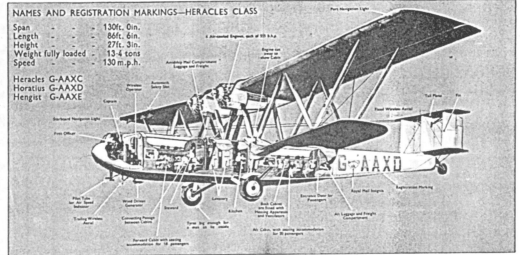

NAMES AND REGISTRATION MARKINGS—HERACLES CLASS

Span	- - -	130ft, 0in.
Length	- - -	86ft, 6in.
Height	- - -	27ft, 3in.
Weight fully loaded	-	13·4 tons
Speed	- - -	130 m.p.h.

Heracles G-AAXC
Horatius G-AAXD
Hengist G-AAXE

This is the *Heracles* in which you will make your journey
—the largest and most comfortable air liner in the World

Why not slip over to Le Touquet one Sunday or for the week-end? It will make the most delightful change. An hour's flight from London, and you will be enjoying all the charm and gaiety of this famous French resort. There is everything there to make you feel you are really on the Continent—Roulette at the Casino, tennis, golf, boating, bathing and dancing

By Imperial Airways, you leave London after lunch on Friday or Saturday, an hour in the air during which you can have tea, and a few minutes later you are at Le Touquet. You can return after tea on Sunday or Monday

By flying over on Sunday morning you have eight hours in Le Touquet and return the same evening. Your dinner on the return journey on the Sunday excursion is included in the fare and served during the flight

You travel in the lap of luxury. Imperial Airways' air liners are the finest and quietest in the world. They are fitted like Pullmans with armchair seats. Plenty of room to move about and chat with your friends, and no more noise than a railway train. There are two lavatories on board and ample luggage accommodation. When you arrive at the aerodrome of Berck, you are conveyed by car, free of charge, to your Hotel in Le Touquet; and throughout the journey you will receive individual help and attention unknown elsewhere

You should try this gay and novel week-end—afterwards you will thank Imperial Airways for making it possible to enjoy so much in so short a time

IMPERIAL AIRWAYS

Bookings and information about Imperial Airways travel from the principal travel agents, or from AIRWAY TERMINUS, VICTORIA, S.W.1, or Imperial Airways Ltd., Airways House, Charles Street, Lower Regent Street, S.W.1. Telephone: VICtoria 2211 (Day and Night). Telegrams: Impairlim, London

TO LE TOUQUET AND BACK FOR £3 : 15 : 0   Including Dinner in the Air

to the Prime Minister of Portugal. Crilly Airways was awarded a mail contract by the Portuguese and the first of four Fokker FXIIs bought by the airline flew to Lisbon in February 1936. Two months later British Airways (Iberia) Ltd bought out Crilly's foreign undertakings. The service was hampered by the refusal of General Franco to permit British aircraft to fly over Spanish territory on their way to Portugal: nor did the airline have an airliner capable of non-stop flight to Lisbon. The Spanish were embroiled in a civil war and Franco was exerting pressure upon the British government to accord diplomatic recognition to his regime.

In May 1936 British Airways moved their base for continental services from Croydon to Gatwick where the old club house had been replaced by a larger terminal building. Gatwick possessed a railway station served by electric trains from Victoria but the airfield itself remained grass covered. British Airways increased its capital to pay for four new DH86s, thereby bringing their fleet number to thirty aircraft of various kinds.

These new aircraft were claimed to be the best equipped in the country with a Lorenz blind approach receiver, Marconi two-way radio sets and direction finders. Additionally they were fitted with de-icing equipment designed to protect the carburettors and airspeed pitot heaters. For this reason, and because salaries were higher, a number of Imperial Airways pilots resigned to apply for a job with British Airways. Despite these improvements Captain Cripps, a veteran pilot from the first World War, noted that the DH86 "being a biplane with struts and flying wires was a sitting duck to pick up ice . . . luckily passengers seemed unaware that they were flying at risk." British Airways were included in the bookings ban imposed by the railway companies in conjunction with Imperial Airways, but business was successfully conducted and heavy loads were usually carried on flights to Sweden.

Meanwhile Imperial Airways continued to fly its former route pattern in Europe plus a seasonal service to le Touquet utilising two HP42 Heracles and the two Short landplanes *Scylla* and *Syrinx*. The other two Heracles had been transferred to Cairo. It was with these four aircraft, and a few old Argosies in reserve, that Imperial Airways tried to compete against both British Airways and foreign competition. From 58 per cent of the traffic to and from Croydon in 1932 their share of European traffic had declined to 42 per cent a few years later. In 1934 the Company had ordered fourteen Armstrong Whitworth Ensigns which were designed to carry 40 passengers at a cruising speed of 200 mph. If these had been delivered as rapidly as the Empire flying-boats and had proved as reliable, much of the criticism directed at Imperial Airways might have been avoided, but as early as January 1936 the government was urging Armstrong Whitworth to give priority to Whitley bombers.

In August 1936 a government committee of inquiry was set up to investigate an unusual affair involving the Permanent Secretary to the Air Ministry, Sir Christopher Bullock. He had reached the highest rank in the Civil Service and approached Geddes with the suggestion that he might be appointed to the Board of Imperial Airways as one of the two government Directors. The Chairman's reaction was unfavourable but Bullock did not give up. He approached Woods Humphery, made a reference to the recent illness of Geddes, and implied that the Managing Director's appointment as Deputy Chairman to himself would prove an excellent working arrangement. Not surprisingly Geddes complained to the Secretary of State, Lord Swinton. When the committee of inquiry reported they declared that Bullock's conduct "was completely at variance with the spirit of the code which clearly precludes a civil servant interlacing public negotiations with the advancement of personal interest." Bullock had to resign and was replaced as Permanent Secretary by Sir Donald Banks.

In March 1937 Imperial Airways pilots flying the DH86 to Cologne complained that, in the winter conditions, the build-up of ice on the airframe and struts posed a serious hazard to safety. The manager of the Company's European Division was Wolley Dod and he flew off on a night service to Cologne to see for himself. The DH86 in which he was travelling encountered a snowstorm and crashed, killing all on board.

A few months later the Lorenz blind landing system, which had been developed by the Germans, was installed at Croydon, British Airways equipped their aircraft to use it but Imperial Airways was in no hurry to follow suit. Familiar to pilots who flew during and for some years after the Second World War as the Standard Beam Approach (SBA), the system enabled a pilot to approach an airport in cloud or darkness, adjusting his course to maintain the extended centreline of the runway. Over his earphones he heard clear dots if he drifted to one side, and if he over-corrected and flew through the desired steady note of the beam, he heard clear dashes. Once established on the centre line the steady note was momentarily interrupted by a distinct signal when the aircraft passed over a marker as he approached the runway, and a different signal near the runway threshold.

The SBA was not a simple solution to runway approaches when the air was turbulent and the wind gusty. These conditions made it extremely difficult to keep the aircraft on the beam which became extremely narrow near the airfield. Accidents were not infrequent. In November one of the British Airways Fokkers, on the night mail service from Sweden, struck the trees bordering the airport on a low approach to Gatwick. Both pilots were killed.

Shortly after this British Airways appointed Alan Campbell Orde to be operations manager, and thus to perform a similar role to that of Brackley for Imperial Airways. Both men had similar backgrounds. During the First World War they had been pilots in the Royal Naval Air Service. Campbell Orde had subsequently been an adviser to the Chinese Air Force when Brackley had been an instructor for the Japanese Naval Air Service. The former had flown for Aircraft Transport and Travel before joining Armstrong Whitworth as their chief test pilot. He had been fortunate to survive the crash of the prototype Atalanta when all four engines had cut owing to a fuel feed problem.

In May 1937 the prototype Ensign was still under development so *Scylla* and *Syrinx* continued to carry passengers to Paris. The Company's ability to maintain its services was further strained by the destruction of the HP42 *Hengist* in a fire in the airship hangar in Karachi. In January 1938 the Armstrong Whitworth Ensign finally made its first flight, but the test pilots were less than happy about the aircraft's flying controls. More hopeful was progress with de Havilland's DH91 Albatross, which had been designed and built in sixteen months. Imperial Airways ordered five and named them the 'Frobisher' class. Geoffrey de Havilland invited Brackley to accompany him on a flight and the latter expressed himself delighted with its handling qualities.

One of seven de Havilland Albatross airliners which entered service in 1938. Originally intended for eastern and transatlantic mail routes their performance was disappointing. The Speedbird emblem on the nose was retained on the successors of Imperial Airways until 1985. *Croydon Airport Society.*

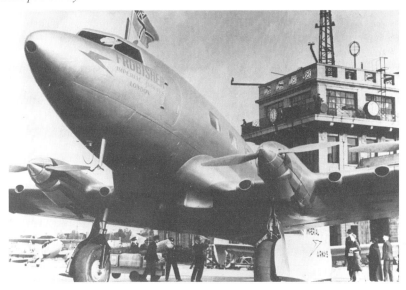

The wet weather in the early months of 1938 reduced the grass surface of Gatwick's airfield to a sea of mud and obliged the airline to move its operations to Heston. At the same time, and by arrangement with the Air Ministry, RAF pilots flew as co-pilots on the flight to Germany conducted by British Airways to make themselves familiar with that country from the air.

In May Lord Swinton was replaced as Secretary of State for Air by Sir Kingsley Wood. Sitting in the House of Commons rather than in the Lords the latter could deal at first hand with the questions and criticisms directed at the conduct of civil aviation. These were growing increasingly vociferous and form the subject of a later chapter.

In June the Ensign was granted a Certificate of Airworthiness subject to a number of modifications, which would take several more months. Meanwhile de Havilland had begun production of a rival to the Lockheed 14 of British Airways. This was not of wooden construction like the DH91, but a twin-engined metal monoplane to be known as the DH95 Flamingo.

In October the Ensign was finally delivered to Imperial Airways. The cabin contained 27 seats and a number of guests were invited to enjoy a flight over London. Some of these observed that the windows were a lot smaller than those of the HP42. In December the Ensigns were assigned to carry the considerable amount of Christmas mail to Australia. Their performance after years of hopeful expectation came as a bitter disappointment to the Company. *Egeria* got no further than Athens before an engine had to be changed. *Elsinore* flew as far as Karachi before an engine failed and *Euterpe* did not reach India. Imperial Airways had accepted nine Ensigns but all were returned to Armstrong Whitworth. More satisfactory was the performance of two DH91 Frobishers which were used to carry the mail to Cairo.

The Ensigns had been found to weigh 20 per cent more than the design estimate, and the original Armstrong Siddeley Tiger engines had to be replaced by American Wright Cyclones. These had nearly twice the horsepower originally intended for the aircraft. Imperial Airways received £42,500 from Armstrong Whitworth in compensation. The Company's problems continued when the Frobishers were grounded following the collapse of the undercarriage, as one of these aircraft was taxying through the mud and slush at Croydon on a windy day.

Despite these vicissitudes the cruising speed of the Ensigns and Frobishers had allowed the scheduled time from London to Paris to be reduced from two and a half hours to an hour and fifteen minutes. The old HP42 had sometimes taken over three hours to make the journey. To quote C. G. Grey, this was bound to happen "when the wind was blowing in the wrong direction."

CHAPTER SEVEN

Australia and the Far East

The Indian government had agreed to the appointment of Lieutenant-Colonel Shelmerdine as Director of Civil Aviation before the arrival of Sir Samuel Hoare in the de Havilland Hercules in 1927. His journey had been intended to inaugurate a regular service from England to Delhi but difficulties with the Persians had prevented this happening. This delay allowed time for the Indian authorities to prepare airfields and areas for flying-boats to alight. Their survey extended beyond Calcutta to Victoria Point in Burma and anticipated an extension of the eventual air service to Rangoon. The RAF also contributed to the future plans of Imperial Airways by carrying out a survey of the route between Calcutta and Singapore.

In October 1928 the Indian government invited tenders for a service between Karachi and Delhi and in April of the following year Indian Airways was formed and affiliated to Imperial Airways. On 28th December a de Havilland Hercules on charter to the Indian company flew the English mails on to Delhi. By November 1930 the airfields at Allahabad, Gaya and Calcutta were prepared for operation in both dry and monsoon seasons.

Imperial Airways had submitted its plans to the British government for a service through to Australia, but the Indian authorities had not decided whether to run their own service across the country or to allow a private company a concession to do so. It had been the intention of Imperial Airways to include Indians on the Board when, as they expected, they were awarded the concession. Then the world financial depression affected India and the death of Sir Sefton Brancker among the passengers on the airship R101 was followed by the-recall to England of Lieutenant-Colonel Shelmerdine to replace him. Plans to continue the air service beyond Delhi were put in abeyance, but the two Hercules airliners on charter were able to continue to bring the mails to the Indian capital until the expiry of their contract at the end of 1931.

This was a bitter blow to Imperial Airways, particularly as India had allowed both KLM and the French Air Orient company to fly across their territory. KLM had begun a fortnightly service to Batavia, in Java, in September 1930. As the financial crisis deepened the Indian Government suspended all expenditure on civil aviation except on wireless telegraphy and meteorology. No subsidies were allowed and the embryo Indian State Air Service expired. By the end of 1931 Karachi was once again the terminus of Imperial Airways' service to the East. Both KLM and the Air Orient company offered to carry on the mail brought by Imperial Airways to Karachi, but the Indian government refused to permit either to do so on the grounds that it was still considering whether to begin its own air service.

KLM benefited because air mail for Singapore was directed from England to Holland. The Dutch airline was also fortunate in that their post office had adopted the maritime practice of paying for an agreed amount of space on their airliners whether or not it was filled. This guaranteed KLM a definite revenue. The GPO did not extend this concession to Imperial Airways until the Empire Air Mail Scheme came into being in 1937.

In April 1931 Imperial Airways, the Air Ministry and the Australian government agreed to run two experimental round trips to carry the air mail between London and Sydney. The Treasury had agreed to finance these flights and provide a subsidy subject to a similar undertaking from other participating authorities. From Croydon an Argosy set off with the air mail for Australia. At Karachi it was transferred to a Hercules commanded by R. Mollard. This had been modified to provide an enclosed cockpit, extra fuel tanks and a luggage breast beneath the fuselage. Southbound from Rambang in Indonesia the aircraft developed a petrol leak and a headwind reduced its speed over the ground to about 70 mph. There was no hope of reaching Darwin and Mollard had to find a flat field for a forced landing. He sighted the racecourse at Koepang and made for a lush green field alongside the track. Unfortunately the long grass concealed large rocks and the Hercules was damaged beyond repair.

Southern Sun Kingsford Smith's aircraft carrying the first airmail to Britain from Australia.

The Australians were waiting for the mail at Darwin and Kingsford Smith's Australian National Airways was ready to carry it on from Brisbane to Sydney and Melbourne. When news of the accident was received Kingsford Smith personally flew to Koepang to pick up the mail and gave Mollard a lift to Darwin. The latter received instructions from London to take a train to Perth and buy another Hercules which West Australia Airways was prepared to sell. There was then no rail link with Perth so Mollard made the journey by sea. Subsequently he flew the Hercules which had been bought and carried the air mail to England.

The experimental air mail flights proved to be expensive and very eventful. Later in the year when Kingsford Smith flew to England he returned to find that Australian National Airways had been obliged to go out of business. But these flights and others by Alan Cobham, Amy Johnson and Bert Hinkler had demonstrated that a service was feasible, although not how much it would cost.

In the spring of 1932 the situation still existed whereby KLM was carrying the British air mails to Singapore while Rangoon in Britain's colony of Burma had no air mail service at all. In a debate in the House of Commons Harold Balfour blamed Imperial Airways for the obstructive tactics of the Indian authorities. Members were reminded of the crass remarks and behaviour of Lord Chetwynd a few years earlier on his visit to India, and they found it unbelievable that Imperial Airways were prevented from flying across India while the Dutch and French did so.

Fortunately, as the financial situation improved, the Indian government became more serious about creating Indian Trans-Continental Airways which would be owned jointly with Imperial Airways. When ready this airline would operate the route from Karachi to Rangoon. In February 1933 a survey by Imperial Airways suggested that it would then be possible to operate from Rangoon to Singapore by way of Bangkok and Penang. Even before the Indian government had named a date when flights east of Karachi would be permitted, Imperial Airways had informed the Australian government that it hoped to connect with an Australian service at Singapore. It was anticipated that the trip from London to Melbourne could be completed in 16 days. Not to be outdone the Dutch subsidiary, which had been operating between Batavia and Melbourne since 1931, offered to run a service from Europe that would take only 14 days.

By May 1933 the Indian government had authorised Indian Trans-Continental Airways to open the Karachi–Rangoon service and Imperial Airways arranged to lease several Atalantas. It was in one of these that at the end of that month Air Superintendent Brackley left Croydon to survey the route from Singapore to Australia. Before he left he had to win an argument with Woods Humphery and Burchall. Brackley's route would take him over

The Air Superintendent, Major Brackley (2nd from left) before the survey flight to Australia of the Atalanta *Astraea* in May 1933. *Croydon Airport Society.*

the Timor Sea and he would face unknown hazards. He wanted extra fuel tanks to be installed in the Atalanta, extra insurance cover for himself, and financial recognition of the risks which he would incur.

He had not over-estimated these risks. Seventy-five miles short of Darwin he realised that insufficient fuel remained to complete the flight. He was approaching Bathurst Island, not knowing whether it was inhabited, when he was astonished to see that it possessed an airstrip and he landed the Atalanta undamaged. The strip had been cleared a few weeks earlier through the initiative and efforts of Father Gsell, a Franciscan priest in charge of a mission to the Aboriginals, none of whom had ever been so close to an aircraft before. Father Gsell had anticipated that some day his airstrip might prove useful. He took care of Brackley and his crew and sent the mission lugger to Darwin for fuel. A few days later the Atalanta was able to fly on to Melbourne.

In Australia, Imperial Airways incurred some unpopularity from a suspicion that the Company was trying to exclude a home-based airline. Two Australian rivals were competing for the route to Singapore. A few months before Brackley had flown from England, Australian Empire Airways was organised to tender for the service to Britain. The Board included Sir Eric Geddes, Woods Humphery and three Australians who had shipping interests. The tenders had to be presented by January 1934, but the range requirements were greater than those of any current British aircraft, whilst the specifications stipulated that the aircraft chosen should be British. In July Imperial Airways opened the route from Karachi to Singapore using Atalantas. One of these was later destroyed by fire in a hangar at Delhi. Tough competition on this route came from KLM which operated Fokkers.

In November 1933 Hudson Fysh, an Australian pilot who had served in the First World War, visited England for discussions with Woods Humphery. Fysh had helped to establish Queensland and Northern Territories Aerial Services in 1920 and had been associated with Imperial Airways in the experimental air mail flights during 1931. Together they approached Geoffrey de Havilland, the aircraft manufacturer who had already established a factory and facilities in Australia. Design work had already begun on the DH86, a much larger four-engined version of the D84. It made its first flight in January 1934 and obtained a Certificate of Airworthiness very shortly before the closing date for the tender. As the DH86 outperformed the specified requirements Imperial Airways placed an order for a number of these for the Australian route. This was followed by the creation of Qantas Empire Airways, a revised version of Australian Empire Airways. QEA and Imperial Airways both nominated Directors and each held 49 per cent of the stock, with the rest placed in the hands of an umpire.

Kingsford Smith was associated with a rival tender for the Singapore service but the award went to QEA. The Australian government agreed to pay a subsidy over a five year period and to establish aerodromes in conjunction with QEA. It was December 1934 before the first scheduled air mail service left Croydon on the 11,000 mile journey to Australia, but two months earlier there had been an air race from Mildenhall to Melbourne for a prize of £10,000 offered by Sir Macpherson Robertson. Although this was won by two Britons, Scott and Campbell Black, in a specially designed racing aircraft, the entry of KLM, a Douglas DC2 passenger aircraft, arrived in second place in a total time of about four days.

Brackley's flying time between the relevant staging posts suggested that a service to Australia occupying five days could have been run. Even so, when the schedule between London and Rangoon was published that journey was going to take eight days. Critics of Imperial Airways' monopoly did not fail to notice the difference between the schedules of the Company and those of KLM. The Dutch flew aircraft which were faster and spent less time on the ground. The places at which their airliners called were chosen as much for servicing and refuelling as for any commercial purpose, because they were restricted on their rights to pick up passengers.

At the tenth annual general meeting of Imperial Airways Sir Eric Geddes agreed that faster aircraft were desirable:

> I need hardly say that we wish to take advantage of the increase in air speed offered by scientific development, but within economic limits represented by government payments for mail and subsidies. There is only a limited field by which we can provide additional speed. Firstly: if the shareholders decide that the Company can be run at a loss. Secondly: if the user will pay more. Thirdly: if the government wants higher speeds they must pay for it. Let us be quite clear that someone must pay the bill because our business is to operate whatever our clients are prepared to pay for ... it is for the government and our customers to say and demonstrate which class and which scale of charges they wish to have.

Range as well as speed had been a problem for the Company because airliners with a greater range could have overflown politically awkward places. Adding extra fuel tanks to the various aircraft merely reduced the payload. From 1932 and for a number of years Imperial Airways urged the Air Ministry to raise half the money required to build a prototype composite aircraft. Major Mayo designed the Mercury seaplane for this purpose. The Air Ministry did finally agree to provide half its cost but refused to pay for a fleet of a dozen seaplanes.

Despite their insistence that the air mail service between Singapore and Melbourne should be run exclusively by Qantas, the Australians were quite unprepared to cope with the rising tide of letters as Christmas 1934 approached. Imperial Airways were fully stretched by the need to double the number of their flights. To help out, Qantas Atalantas were used between Singapore and Darwin. An Atalanta, which tried to make up time by flying over the Timor Sea at night, had to put down in the desert when the airport authority at Darwin failed to receive notification of its arrival and had not prepared a flarepath. Although five DH86s had been flown out to Qantas by the end of January 1935, others were held up by the refusal of the Australian government to issue the type a Certificate of Airworthiness, until a modification which they had demanded was carried out.

One of the Atalantas was still operating to Darwin when in April 1935 the first passengers left Croydon to fly through to Brisbane. In Australia the demand for air mail space was so heavy that the DH86s had to be found extra landing grounds as fuel had to be reduced to avoid overloading the aircraft. In September Imperial Airways doubled the number of services south from Calcutta, but Qantas were in no position to handle the extra loads and the Company had again to offer assistance.

Kingsford Smith tried to re-enter the airline business with a proposal to operate American Sikorsky flying-boats across the Tasman Sea to New Zealand. The New Zealand government turned this down as negotiations were already in progress with Imperial Airways. The British government was willing to pay half the cost of the service to New Zealand and made it clear

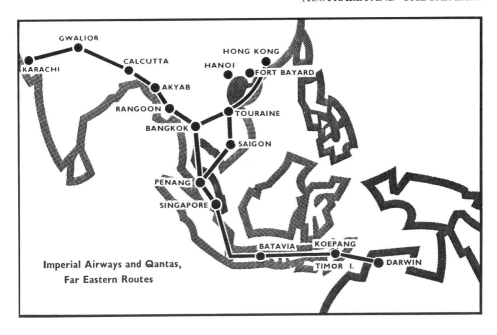

Imperial Airways and Qantas,
Far Eastern Routes

that it would go ahead, with or without Australian participation. The New Zealand government was also adamant that it would run its own service or use that of Imperial Airways. Kingsford-Smith's pioneering exploits had long before made him a hero to the Australians and, once again, Imperial Airways was a target for a perceived offence to national honour; but when it became clear that some of their politicians had used Kingsford-Smith's name for their own domestic purposes the matter was seen in a different light.

At the end of 1935 the Company suffered several accidents which put a strain on their schedules. On 2nd November the Atalanta class *Astraea* overshot the landing area at Rangoon. The pilot was only slightly injured and repairs to the aircraft were undertaken locally. Next day *Atalanta* hit a tree on a pre-dawn take off from Kisumu. The crew were injured but none of the passengers were hurt. They included Lord Balfour of Burleigh. During the month Sir Charles Kingsford Smith tried to beat the record for the fastest flight from England to Australia, flying a Lockheed Altair. His aircraft was seen over the Bay of Bengal but no further trace of it was found.

Misfortune overtook two flying-boats before the end of the year. The Short Kent *Sylvanus* was burnt beyond repair when it caught fire during refuelling at Brindisi. As a result of this one of the old Calcuttas had to be returned to service. This aircraft, *City of Khartoum*, was approaching

85

Sir Charles Kingsford Smith.

Alexandria in sight of the flarepath crew when all three engines stopped, and she dropped into the sea outside the harbour. The launch used by the flarepath crew had no radio, nor was it sufficiently seaworthy to venture beyond the harbour to attempt a rescue. Several hours passed before HMS *Beagle* was able to find the sole survivor, the pilot.

The inquiry speedily established that lack of fuel was the cause of the accident. The fuel gauges on the Calcutta class were known to be inaccurate but the situation had been made worse by the incorrect adjustment of the fuel jets in the carburettors. This had the effect of increasing consumption by ten per cent. Enquiries revealed that the Calcuttas on the lengthy Mirabella–Alexandria sector often arrived with little fuel remaining. On the night of this particular disaster *City of Stonehaven* arrived with fuel for only another twelve minutes flying in her tanks.

Air Superintendent Brackley had always insisted that the pilot in command should decide for himself how much fuel he considered necessary for each flight. If a sector could not be completed safely due to an unexpected headwind he should turn back. One flaw in such flight planning arises when the increase in wind strength is noticed after the aircraft has flown past the point of no return. Moreover, even with full tanks, the Calcuttas seldom had much fuel left when completing this sector and the load carried did not always allow the fuel tanks to be filled to capacity. In March 1933 this same pilot had landed twenty miles short of Athens on a flight from Alexandria. He was transferred to landplanes for a time, and had then landed short of Darwin on a mail flight to Australia. This had also happened to the pilot R. Mollard in a Hercules and to the Air Superintendent himself in an Atalanta. Air Ministry regulations had never required airline operators to carry a legally defined minimum fuel reserve, as was the case in the United States: nor did the Company take such action on its own initiative after this accident. Imperial Airways did however increase the tank capacity of the Calcuttas by fifty gallons. Unaccountably no steps were taken to equip aircraft with life-rafts.

Despite these misfortunes the Company managed to maintain its schedule to Singapore and aircraft never took more than one day longer than the eight days published in the timetable. In 1935 only 410 passengers from England flew as far as India or further east, a number which could be accommodated on one Boeing 747 today.

In August 1936 the Short Kent *Scipio* made a heavy landing and sank at

Short Kent *Scipio* flying boat which was superseded by the Short Empire boats.

Mirabella. The wireless operator had inadvertently moved the tailplane trim to full nose down. On his approach the pilot realised something was wrong and opened the throttles to climb away. This compounded the problem and the aircraft dived into the sea. Within a week the HP42 *Horsa* tried to make up a delayed schedule by flying on to Bahrein for a night landing. Unfortunately the signal informing that station was not received, no flarepath had been prepared, and in the darkness *Horsa* flew past Bahrein and eventually landed in the desert 100 miles away. A brief message had been transmitted shortly prior to touchdown and the following day the RAF found their landing place and recovered the passengers.

Imperial Airways did not begin a regular scheduled service to Hong Kong until 1936. One reason for the delay was the attitude of the Indian government which, until 1933, had prevented the Company from running a commercial service over their territory. Another reason was the perennial one of shortage of aircraft. Then Imperial Airways reacted to the enterprise of both Germany's Luft Hansa and the French Air Orient Company. The former was operating from Berlin to Nanking and Shanghai whilst the latter, whose Far East route had terminated in Saigon, now extended their route to take in Canton and Hong Kong. Pan American Airways were also able to fly to China through affiliation with the China National Aviation Company. Until Imperial Airways established their own service to Hong Kong, the Americans had to be content with the Portuguese island of Macao as a terminus in the Far East.

By August 1935 a new company, Imperial Airways (Far East) Ltd, was registered and on 16th September a DH86 left England for Penang to conduct survey flights. On the first run from England carrying mail Captain Armstrong flew from Penang to Hong Kong in one day, a distance of 1,852 miles, returning to Penang on the following day. Regular scheduled services began on 23rd March 1936 enabling mail from England to reach Hong Kong in ten days. The journey by sea took 34 days. In March 1937 the service was subsidised at the rate of £25,500 per year and in the autumn of that year two more DH86s were sent out to cope with the increasing amount of air mail. An agreement with Siam permitted the shortening of the route and a reduction of the London–Hong Kong schedule to eight and a half days. In November 1938 the subsidy was increased to £56,000 which included grants from the governments of Malaya and Hong Kong. DH86s continued to be used because Imperial Airways possessed insufficient flying-boats to put any on this route. One of the DH86s was destroyed by fire at Bangkok in 1938 but the service continued to operate throughout the lifetime of Imperial Airways.

CHAPTER EIGHT

The South Atlantic, Portugal and West Africa

When the colonial subjects of the Spanish Empire were fighting for their independence in the early 19th Century the United Kingdom gave them every encouragement and welcomed the opportunity to take advantage of the opening of trade. Great Britain was the 'workshop of the world' and the South American continent provided a huge market for the products of the manufacturers. Investment in South America included the creation and running of a vast network of railways. The United Kingdom was still South America's principal European trading partner at the end of the First World War, and at least half a dozen shipping companies ran cargo and passenger services to that continent from London, Liverpool and other ports.

With the development of civil aviation it was to be expected that efforts would be made to link Europe by air to South America. Sadly the lead was not taken by British mercantile concerns and very little encouragement was offered by the British government. In 1919 the Handley Page Company expressed an interest but nothing happened.

In 1922 the first east to west crossing of the South Atlantic ocean was made by two Portuguese airmen flying a Fairey IIID seaplane from Lisbon. *En route* they landed at Las Palmas in the Canary Islands, St. Vincent in the Cape Verde Islands, Santiago Island and the tiny St. Paul Island, literally a collection of rocks in the middle of the South Atlantic. There a Portuguese cruiser awaited them, but their seaplane ran out of fuel just short of St. Paul and was damaged beyond repair when it landed in the ocean. The two airmen were rescued unhurt and a second seaplane was shipped out to them, then a third, as further mishaps dogged their endeavour. Eventually they completed their flight to Rio de Janeiro.

In the following years other adventurous Spanish and Italian airmen managed the crossing of the South Atlantic, and in 1927 two Frenchmen made the first non-stop flight from Senegal to Brazil in 18 hours. It was not long before the commercial possibilities were exploited. In 1924 the French company Aéropostale had started an air mail service from Toulouse to Dakar in Senegal. From this busy port mail was carried by sea to South America. In successive years the GPO paid increased sums to the French to carry British mail. In 1928 it was £60,000. The amounts grew as the Germans also introduced a combined air and sea mail service to South America. These

Laté 300. The Latecoere company's aircraft designed for the Senegal-Brazil route made the first South Atlantic crossing in 18 hours 50 minutes in 1934. Piloted by Jean Mermoz one of these was lost in the South Atlantic in Feb 1936. *John Stroud.*

payments would have formed a useful subsidy to a British airline if sufficient encouragement had been offered but the Air Ministry and the government refused any financial support. By May 1930 Aéropostale had aircraft capable of flying the mail from Senegal across the South Atlantic to Pernambuco in Brazil. Thereafter it was flown south to the Argentine and to Chile by the French airline.

Like the French the Germans had found themselves excluded from the opportunity to develop an air route across the North Atlantic by the denial of rights to use landing places in Bermuda or the Azores, Ireland or Newfoundland. Undeterred, the Germans experimented with various schemes to inaugurate a successful air mail service in competition with the French to South America. No such enterprise was shown by Britain. The Air Ministry allowed Vickers to begin work on a large six-engined monoplane flying-boat which was designed for long ocean crossings. Then in 1932 the project was cancelled before the first machine was completed. The continuing financial depression and the prior claims of the Empire routes served as an explanation, but another reason was the Air Ministry's reluctance to allow spending on anything which was not of equal benefit to the RAF.

The more successful of the two separate services which were introduced by the Germans was the *Graf Zeppelin* airship schedule between Berlin and Rio de Janeiro. In 1932 and in 1933 nine flights were made to Brazil, and there were twelve in 1934, when 920 passengers were carried along with 31,000 lbs of freight and mail. Any possibility that Britain might compete with an airship of her own had been ruled out by the disaster to the R101 in 1930. Although the R100 built by a subsidiary of Vickers had made a successful crossing of the North Atlantic the government instantly abandoned the airship programme.

The other German air service to South America involved the use of a floating base in the middle of the South Atlantic. After a series of trials which had begun in 1932 the first scheduled Luft Hansa mail flight set off from Berlin in February 1934. A Heinkel with 107 lbs of mail flew to Seville via Stuttgart and Marseille. The mail was transferred to a Junkers Ju52 which made a refuelling stop at Las Palmas before continuing to Bathurst in the British colony of Gambia. The mail was then placed on board a Dornier Wal flying-boat which was positioned on a converted steamer, the *Westfalen*. This vessel then sailed 940 miles out into the Atlantic, whereupon the flying-boat was catapulted into the air and flew on to Natal in Brazil. Another Luft Hansa aircraft carried the mail south to Rio de Janeiro and Buenos Aires. In 1934 the airline flew 47 scheduled flights to and from South America. A second catapult ship was put into service and in 1935 a night service was introduced, which enabled mail to be carried between Germany and Rio de Janeiro in three days. In 1936 the Dornier Wals were replaced by Dornier Do18E aircraft. The crews and mechanics had to live in primitive accommodation in Bathurst where many succumbed to malaria and yellow fever. On the *Westfalen* living conditions were extremely cramped.

By 1933 Aéropostale had fallen victim to the financial depression and went into liquidation. Its successor, Air France, was able to introduce the Latécoère 301 flying-boat and three other types of aircraft on the South

In 1934 Luft Hansa opened a service to South America. The Heinkel He 70 (left) flew to North Africa transferring its cargo to a Junkers 52 (right) which continued to Bathurst, Gambia. A Dornier Wal seaplane on the ship *Westfalen* was launched in the South Atlantic and flew to Natal, Brazil. *Robert Jackson*.

American route in 1936. During that year these machines made 86 crossings of the South Atlantic.

The Air Ministry was particularly interested in the success of the German catapult floating bases but came to the conclusion that the high acceleration necessary would not be tolerated by passengers. As the United States was known to be preparing a combined mail and passenger air service the catapult idea was not thought to be worth pursuing. In 1935 the Under-Secretary of State for Air told the House of Commons that the Air Ministry would help an airline which wanted to establish a route to South America, but no financial assistance could be offered by the government.

In March 1936 the government committee chaired by Sir Warren Fisher recommended financial support for a second British airline, which would thereby allow Imperial Airways to concentrate on routes to the Empire. The committee also considered subsidising an air service to South America and tenders were invited and examined. But the government continued to insist upon the use of British aircraft with British engines. This was a totally unrealistic condition. Faced with a resurgent Germany, re-armament had begun and no British aircraft manufacturer was in a position to produce a suitable airliner for a long ocean crossing, whilst the Air Ministry was demanding that military orders had to be accorded the highest priority.

In February 1937 the government declared that British Airways Ltd had been chosen to operate the route to Bathurst, and it was the intention that the airline would thereafter extend the service from West Africa to South America. In July the government agreed that British Airways would require financial support to establish the route and two months later discussions began with the Portuguese authorities about the use of Lisbon as a stop.

All of this was good news for British merchant houses in South America. In Britain the GPO had paid £87,500 to the French and Germans for carrying the mail in 1936 and would pay £98,000 in 1937. In June 1938 British Airways sent a group to survey the route between Brazil and the Argentine, whilst another team was busily negotiating with foreign governments for permits to fly over their territories. In September British Airways received their first Lockheed 14 which had arrived in crates at Southampton, was re-assembled and flown to Heston. The following month it was flown to Lisbon via Bordeaux, covering just over 1,000 miles in under seven hours.

In December the House of Commons was informed that from January 1939 a direct London to Lisbon service would begin, and that British Airways would be paid a subsidy of £116,000 to operate it. The service did not begin in January, nor in March when General Franco was accorded the diplomatic recognition which he had craved for his regime. Given the range of a Lockheed 14 when carrying a worthwhile payload the Spanish rebuff proved a severe setback. Nevertheless a Lockheed made one proving flight

A Dornier Wal is launched from the *Westfalen* in the South Atlantic to carry the mail to Brazil. Feb 1934. *Lufthansa.*

to Bathurst, reaching Lisbon in five hours before flying on to Casablanca, Agadir, Port Etienne and Dakar. Apart from the difficulties with Franco no decision had been taken about a suitable aircraft for the South Atlantic ocean crossing. The original proposed date of 1940 could not possibly be met and the British government admitted that 1943 would be a more realistic target. Until then the GPO would continue to pay the French and Germans to carry the mail. In 1939 the cost for using their services had risen to £118,000. Any hope of a British service ended when war broke out in 1939.

When France capitulated in 1940 the need for a route to West Africa became very important. In November 1934 Woods Humphery had let it be known that Imperial Airways was considering a landplane service westward from Khartoum to link up with the British colonies of Gambia and Sierra Leone. In May 1935 it was decided to survey a route from Khartoum to El Fasher, Fort Lamy, Maidugari, Kano, Kaduna and Lagos, capital of Nigeria. This news was particularly welcome to the South African government, which was becoming increasingly concerned about Mussolini's ambitions in Ethiopia and elsewhere in Africa, which might endanger her trade links with Europe.

It was February 1936 before a DH86 arrived at Khartoum to begin the survey, and it soon transpired that there were no airfields west of Kano in the interior of Africa in any fit state for a scheduled commercial service. The problem in preparing aerodromes was the lack of roads and in the African bush even the camels had difficulty negotiating the existing tracks. Kano was two days' flying time from Khartoum and this was the first stage to be inaugurated. Early in September it was possible to extend the route to Lagos via Kaduna, Minna and Oshogbo. In October the GPO was able to announce that in future air mail posted to Lagos would take only six days to arrive. This was a great improvement on the 19 days which had been customary when the mails had been shipped from England by sea. It was necessary for flying to be carried out during the hours of daylight because no wireless direction-finding stations existed. On each of the two days it took to reach Kano 1,000 miles had to be flown. This made timekeeping very important, particularly on the eastbound journey to Khartoum where there was two hours' less daylight time. At both Kano and El Fasher departures were scheduled at 5.30 a.m. to complete the flight before nightfall. The DH86 carried five passengers in addition to the air mail. In October 1936 the route was extended from Lagos to Accra on the Gold Coast.

The route was operated in conjunction with the Elder Dempster shipping company which was keen to link Bathurst and Freetown by an air service. A Short Scion seaplane was obtained and a pilot engaged to fly this sector, which was opened in June 1938. Then in December a routine maintenance inspection revealed corrosion from salt water to parts of the seaplane. Weeks passed as a repair team travelled out from England, and the delay was exacerbated by the team's failure to have included welding equipment in their baggage.

When the seaplane was finally repaired it was possible to resume the weekly flight via Conakry and Boulama. By connecting with the French and German air services to West Africa mail from England could reach Sierra Leone in six days. In May 1939 the seaplane began a regular service to Takoradi but in August it was struck by flotsam when at its moorings and severely damaged. The war broke out before repairs could be undertaken and the coastal seaplane service was not resumed.

CHAPTER NINE

The Empire Flying-boats

Revenue from the Empire Air Mail Scheme was essential to the future prosperity and development of the international routes of Imperial Airways. The Company proposed to operate these routes with flying-boats and the agreement of the British government to adopt the EAMS was facilitated by the promise that the use of flying-boats would avoid the huge expense of enlarging the existing aerodromes, many of which were often unserviceable during the monsoon seasons. Imperial Airways also needed to compete more effectively against airlines such as KLM whose Douglas DC2s were so much faster than the British landplanes. The Air Ministry agreed to provide the marine facilities.

Short Brothers were asked to tender a design to meet the Company's requirements and offered the 'C' class all-metal high-wing monoplane. It had two decks, the upper one for the crew, the lower for 24 passengers, two stewards, luggage and mail. The passenger area was divided into three cabins. The forward one was a promenade cabin where passengers could stand and look out of the windows. Cruising speed was 165 mph and the original range specification was for 800 miles. The design involved numerous innovations and Shorts wanted to build a prototype, but the demand for seats and cargo space was stretching the Company's resources to the limit. Consequently Imperial Airways ordered 28 machines off the drawing board. A smaller number were ordered by Qantas Empire Airways and the New Zealand airline Tasman.

The first of the new flying-boats to go down its slipway and be taken up by Short's test pilot was *Canopus* on 4th July 1936. By September it had completed its full load trials and went into the workshops to be fitted and furnished for passenger service. *Caledonia* made its first flight on 11th September and four days later began its Certificate of Airworthiness trials. Hardly any time was spent on proving flights and, as the aircraft were delivered to Imperial Airways at the rate of one a month, some of them were accepted with as little as one flight recorded in their log book.

The Company began advertising the Empire class as the fastest flying-boats in the world, which would carry passengers to India in five days, Australia in ten days. On 22nd October Major Brackley flew *Canopus* to Rome via Bordeaux and Marseille. Waiting to fly her on to Alexandria was J. F. Bailey, one of the original pilots who had joined from the British Marine

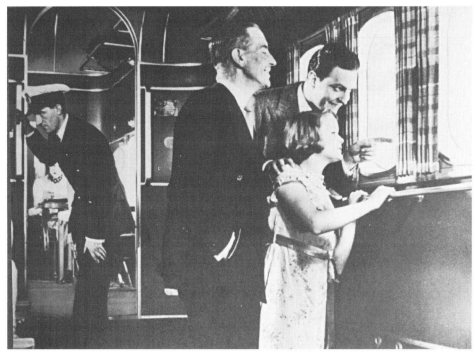

View of the observation deck of an Empire flying boat.

Air Navigation Company. From Alexandria he flew *Canopus* back to Brindisi. *Centaurus* was delivered by Shorts in October, *Cavalier* in November and *Castor* in December. Egypt then joined the EAMS and that month *Centaurus* flew out with the first of the mail to earn the Egyptian subsidy of £22,500.

Henceforth all letters between countries of the British Empire could be sent by air at a penny ha'penny per half-ounce. It had been intended to send *Caledonia* on a trial run across the North Atlantic, but Newfoundland's waters were frozen over, so she was flown to Alexandria with five tons of Christmas mail. From there she continued to India for route proving. On the way home *Caledonia* covered the distance from Alexandria to Marseille in just over eleven hours, and thereafter to Southampton in four hours. In January 1937 *Castor* and *Centaurus* initiated regular service flights as far as Alexandria. Passengers embarked at Marseille and stops were made at Rome (Lake Bracciano), Brindisi and Athens. In February *Caledonia*, as part of her preparation for her trials on the North Atlantic, flew the 2,200 miles from Southampton to Alexandria in thirteen and a half hours.

During March the last landplane service from Africa landed at Croydon and all future outbound flights began from Hythe, Hampshire. Services to

Africa departed on Tuesdays and Fridays and to Australia on Wednesdays. Scheduled times were reduced, eastbound passengers spending the second night out from England in Cairo. In May *Cambria* was flown to Khartoum and at the end of June *Centurion* inaugurated the first EAMS flight to South Africa. Thereafter the mails were carried all the way to Durban by flying-boats, travelling with three tons at a time. As winter approached the weight of mail steadily increased and *Cambria* and *Caledonia* were required to carry five tons. Even so a backlog grew in London, whilst South African Airways was hard pressed to cope with the arriving loads. By mid-1938 a third service each week was scheduled to Durban.

The conversion to flying-boats from landplanes of most of the Company's pilots and the training of numerous new recruits was a major task. Initially the old Calcuttas were used until a sufficient number of Empire flying-boats was available. The pilots had to be introduced to the basic principles of watermanship, how to approach a mooring buoy sufficiently slowly when the wind was blowing in one direction and the tide flowing in another. They had always been required to know the Morse Code but now semaphore had to be mastered too. Marine law had to be studied including a knowledge of the lights and signals displayed by vessels, both by day and by night. A master, as a flying-boat captain was also known, who landed some distance from his destination owing to fog, and gladly accepted a tow into harbour from an obliging tugboat skipper, learned soon afterwards that the Company had received a claim for salvage.

Nevertheless the pilots of the Empire boats were rightly proud to be flying such modern aircraft. Imperial Airways encouraged their captains to leave the cockpit at suitable times and to converse with the passengers. Most of the latter, except for the very nervous, appreciated an opportunity to talk to their pilot. The crew usually stayed at the same hotels as the passengers on nightstops on international routes, so they might spend a number of consecutive days in each other's company. In India and the Far East some of these hotels expected their guests to dine in evening dress. A command on an Empire boat was some consolation for the pilot's small salary and the requirement to pack a dinner jacket.

In June, in preparation for the extension to Singapore of Empire boat services, the old Kent class *Satyrus* left Alexandria on a survey flight, landing at eighteen different locations which provided marine facilities before reaching Singapore. *Ceres* made a trial run in September, landing in the Dead Sea owing to religious objections to the use of Lake Galilee. In October *Calypso* opened the service to Karachi and after another survey flight by *Satyrus* it only remained for sufficient new aircraft to be delivered to open the route to Singapore. By the end of 1937 Imperial Airways had taken delivery of 22 Empire flying-boats.

This and other extensions were delayed by a series of accidents. In March 1937 *Capricornus* had left England with the first through mail to Australia. Running into snowstorms over the French Alps the pilot attempted to turn back, and informed Lyons Airport that he was not sure of his position. The aircraft then fell victim to icing and crashed into a mountain. The sole survivor was the wireless operator, whose injuries did not prevent him from finding a farmhouse and reporting the disaster.

The following month, April 1937, *Courtier* crashed in Phaleron Bay off Athens. On his approach a newly promoted captain misjudged his height over a flat calm sea which blended with the sky, and dropped the flying-boat so heavily that she split open. Three of the passengers were drowned, others injured. It was after this accident that the Air Ministry made the provision of lap straps in airliners compulsory. In May *Corsair* was blown off her moorings in a gale and jammed her tail in a yacht, whilst *Castor* was rammed by a launch at Hythe, but neither of these aircraft was out of service for very long.

In December the very experienced Captain Mollard inadvertently moved the flap selector fully down, before attempting to take off in *Cygnus* from the harbour at Brindisi. In choppy seas the flying boat lifted off at too slow a speed, stalled in and porpoised severely before nosing in and smashing her bow. Two of those on board were drowned, and all the others were injured including Sir John Salmond, one of the government Directors on the Board of Imperial Airways. The first officer had been thrown into the sea but swam back to pull three passengers from the hull. Subsequently he was awarded the Royal Humane Society's Stanhope Medal. When the Report on this accident was published the Company was censured for failing to have specified sufficient push-out windows and escape hatches when ordering the flying-boats. It so happened that a Member of Parliament had eight months earlier put down a question, asking why only small roof hatches had been fitted, but it required this accident for modifications to be carried out.

By February 1938 there were sufficient Empire boats in service to open the route to Singapore, and during the summer floating landing stages off the Southern Railway's docks at Southampton were completed, allowing flying-boats to be loaded and unloaded more swiftly. A new electric flarepath had also been developed by Imperial Airways for night take-offs.

Mishaps during the year 1938 included damage to a flying-boat when it was towed into an Italian submarine in Naples harbour. In a separate incident *Ceres* landed on Lake Dingari in India and then got stuck in the mud. The landing was unscheduled and the nearest telephone was twenty miles away. On 27th November *Calpurnia* was destroyed at Habbaniyah when a dust storm blew up as the pilot was trying to land on the lake. The captain and three crew members were killed. The flying-boat was carrying Christmas mail and no fare paying passengers were being carried. In January 1939

Imperial Airways Short flying boat *Caledonia* over Manhattan in 1937. The fuel capacity ruled out any ability to carry a payload.

Calypso developed engine trouble when flying off the coast of Normandy. The pilot landed successfully on the open sea and a tug towed the flying-boat into the harbour at Cherbourg. Those on board were luckier than the crew and passengers of *Cavalier*, which came down between New York and Bermuda during that month, owing to carburettor icing. This accident and its aftermath are covered more fully in the next chapter.

On 14th March 1939 *Corsair* was running short of fuel over the Belgian Congo and her pilot put her down on the shallow River Dangu. The hull was damaged and soon filled with water. The nearest airfield was at Juba, 150 miles away. A rescue team comprising engineers from Imperial Airways and Shorts flew out to recover the crew and passengers and to make repairs. Their first task was to enlist the help of local natives, to hack through areas of jungle to reach the river. With their help they beached *Corsair* and began the repairs, constantly plagued by mosquitoes and obliged to endure the intense heat.

They were still engaged on this task during May when *Challenger* crashed in Mozambique. The pilot had not made a circuit and attempted to land in the opposite direction to his only previous one. He failed to see a small jetty and the flying-boat ricocheted off it into very shallow water. In June *Centurion* crashed in the Hooghly River near Calcutta, and *Connemara* was burnt out when a refuelling barge caught fire at Hythe. By the end of that month nine of the Empire boats had been written off within two years of the delivery of the first one. It had been intended to send both *Champion* and *Clyde* to Bermuda to replace the lost *Cavalier* but these accidents so depleted the Company's fleet that no aircraft could be spared to resume the service to New York.

Meanwhile the repairs to *Corsair* were completed and Captain Kelly Rogers was sent out to see if it would be possible to fly her out. The seasonal rains had raised the level of the river but there was an awkward bend along the stretch of water required for the take-off run. Kelly Rogers made the attempt, but the flying-boat lost speed as he tried to negotiate the bend, and he abandoned his run. Taxying back *Corsair* struck a submerged rock which holed her hull, allowing water to flood in. Drums were lashed to the flying-boat to lift her, the engines removed to reduce her weight, and repairs were begun again. This time the local natives were employed to build a dam and create an artificial lake. Months passed but the dam was completed by the onset of the next rains.

On 6th January 1940 Kelly Rogers was faced with a second attempt to recover *Corsair*. The stretch of water available for the take-off run was only twelve yards wider than the aircraft's wingspan. This time he was successful and after landing at Juba for fuel he continued to Alexandria, where *Corsair* was given a thorough overhaul. As a result of the ten month period which the flying-boat spent on the River Dangu a new village community grew there. It was (and is) called Corsairville The artificial lake is still there, and the village depends for supplies on the road which the rescue team and their helpers hacked through the bush to reach the aircraft.

The Australian government needed some persuading that flying-boats were preferable to landplanes as a means of linking the member nations of the British Commonwealth. Hudson Fysh, Managing Director of Qantas Empire Airways, recognised at an early stage the advantages to be gained. New heavier landplanes would involve a great deal of extra expense on the existing aerodromes in Australia, whilst both Imperial Airways and Pan American proposed to extend their routes to Australia using flying-boats. The route between Darwin and Singapore which these would use was several hundred miles shorter than the landplane route, allowing economies in fuel expenditure. There was also the advantage of flying-boat bases for defence purposes at a time when Japan was posing an increasing military threat. Hudson Fysh accompanied Brackley when they surveyed the route south from Singapore in an RAF flying-boat. Within a month they had selected most of the marine stations which they expected the Empire boats would need to use.

It took until September 1936 for the Australian government to accept the EAMS and their reservations on its implementation were not accepted by the British government until the following February. The Australians did not begin their survey for the first nine bases until October, and in April 1938 the Federal government had only just received from the government of New South Wales the land for the base at Rose Bay, Sydney.

New Zealand showed far more enthusiasm for an air link with England and in August 1937 the two governments agreed that a joint operating company should provide a trans-Tasman service. In a tripartite arrangement, Britain and New Zealand each had a 38.5 per cent share in subsidies and revenue, with the Australian government taking 23 per cent. When *Centaurus* arrived at Auckland from Sydney 50,000 people turned out to watch. In September 1938 it was agreed that Tasman Empire Airways should receive five of the Empire flying-boats, whose range had been increased to 2,000 miles. They received three of these the following year, *Aotearoa*, *Awarna* and *Australia*, the war preventing the despatch of the others.

Australia welcomed the first Empire boats in July 1938 on the arrival from England of *Camilla* and *Capella*. The passengers included seven journalists who were the invited guests of Imperial Airways. They were anything but impressed by their reception at Darwin. Qantas, responsible for handling arrivals, had not taken delivery of a launch. The passengers had to remain on board until the arrival of the Customs officials, and they became increasingly sea-sick in the heavy swell. After that the Port Health officials insisted in vaccinating them. The journalists were able to express their disgust at this welcome in print, and when the next flying-boat arrived all the passengers were taken at once to a local hotel where the necessary formalities were carried out more efficiently. The base at Darwin was also unprepared for bad weather. In December *Coorong*, one of the Empire boats operated by Qantas, foundered in a gale. Later she was salvaged and returned to Imperial Airways.

For a return fare of £274 from London to Sydney the flight was supposed to take nine days, but the Company had difficulties keeping to the published timetable over a route that was 13,000 miles long. The weather, ever increasing consignments of mail and a shortage of aircraft were contributing factors. Aircraft sometimes arrived almost a week late. It was also difficult to turn aircraft round at Sydney fully serviceable for the flight home. A year after the flying-boat service had begun the Australian authorities had still not completed any hangars at Rose Bay.

CHAPTER TEN

The North Atlantic Challenge

In 1919, after Captain John Alcock and Lieutenant Arthur Whitten-Brown had flown the Vickers Vimy from Newfoundland across the ocean to Galway in Ireland, they not only won Lord Northcliffe's prize of £10,000 but were also rewarded with knighthoods. If only for a short time the imagination of the British public was stirred by the future possibilities of long distance travel by air. This was a different sort of triumph from the capture of a few hundred yards of Flanders mud at a cost of tens of thousands of lives.

The first Controller-General of Civil Aviation was Sir Frederick Sykes. He was looking forward to the development of an air route to North America via the Azores and Bermuda and obtained permission from Portugal to make surveys. His plans came to a full stop when the Treasury was unwilling to allocate more than £5,000 towards this venture.

Airships made a few flights across the Atlantic ocean. The British R34 flew to the USA and back in 1919, and in 1924 the German airship LZ126, commanded by Hugo Eckener, flew from Friedrichshafen to Lakehurst, New Jersey. This was a one-way journey, the airship being delivered to the US Navy in part payment of war reparations. A number of attempts to fly the Atlantic in an aeroplane were made by adventurers who paid for their enthusiasm with their lives. But a prize of $25,000 had been offered for the first non-stop flight between New York and Paris, and on 21st May 1927 Charles Lindbergh won it, flying a Ryan monoplane in a time of 33 hours 30 minutes. Two weeks later Clarence Chamberlin successfully flew an even greater distance, in a Bellanca monoplane, New York to Eisleben in Germany, 3905 miles in 43 hours 49 minutes. He did not have quite enough fuel remaining to reach Berlin.

These flights were made at a time when there was no adequate forecasting of weather, nor were there aids to long range navigation such as wireless sets. Blind-flying instrumentation was in its infancy and icing could cripple an aircraft. It was certainly understood that an east to west flight into the prevailing wind would take longer, but that did not prevent several airmen and their sponsors from making the attempt, including the two Imperial Airways pilots, Minchin and Hinchliffe, who lost their lives.

In the mid 1920s the British government believed that airships had a useful role to play and authorised the construction of two. A subsidiary of

Vickers called the Airship Guarantee Company built the R100. The other was constructed by the Air Ministry's airship establishment at Cardington and was known as the R101. This scheme pitted private enterprise against the government's factory, with the advantage to the latter, which had the right to inspect every stage of its rivals preparations. Germany was ahead in airship technology. In October 1928 the *Graf Zeppelin* with 37 crew, 20 passengers and a load of mail flew from Friedrichshafen, passed over Barcelona and Madeira, and crossed United States territory over Chesapeake Bay, circling over Washington and finally mooring at Lakehurst, New Jersey.

During the following year the Germans equipped their ocean liners *Bremen* and *Europa* with a catapult to launch a seaplane carrying mail, when steaming to within 500 miles of their port of destination. The French had adopted this technique with the *Ile de France*. Aéropostale, the French airline, also negotiated exclusive use of the Azores from Portugal, but very soon afterwards the agreement lapsed when the company was overtaken by severe financial problems.

The Wall Street crash of 1929 was followed by a financial depression throughout the world. The British government's policy as regards civil

On the first ocean crossing of Germany's *Bremen* in July 1929 a Heinkel HE 12 seaplane was catapaulted to speed the mail to New York. The launch was made about 300 miles off-shore. *Lufthansa.*

aviation was to give priority to routes to the Empire. Canada was separated from the United Kingdom by so great a distance that the Air Ministry was in no hurry to allocate funds to develop an airliner capable of making the crossing of the North Atlantic. Woods Humphery tried to impress upon Lord Amulree, Secretary of State, the need for the Air Ministry to include one or two designs in its experimental programme. Soon afterwards Vickers began work on a six-engined monoplane flying-boat.

Woods Humphery also travelled to New York to meet the founder of Pan American Airways, Juan Trippe, and to discuss with him the possibility of future airline services carrying passengers and mail between their two countries. Initially they considered a joint company to fly between New York and Bermuda. There was then no aerodrome on the island, where not even the governor was permitted to import a motor car to travel along the narrow streets. At this early stage Woods Humphery and Trippe could do no more than come to an informal understanding, that neither company would operate to the territory of the other, until both were ready to begin a service.

In 1930 Trippe sent Lindbergh to report on the feasibility of the Arctic route, although this would not be an option available to flying-boats in the winter. Trippe had discussed with Woods Humphery an air mail service to Britain, with Pan American flying the mail to Bermuda and Imperial Airways carrying it on via the Azores. But the problem on this southerly route was the vast expanse of ocean separating Bermuda from the Azores. Moreover the US Post Office swiftly declared that US mail services must be the sole responsibility of US citizens. It invited bids for a trans-Atlantic air mail service. One requirement was that 300 lbs must be carried each way in a fire-proof mail compartment at a rate of two dollars a mile. There were no bids.

Germany, France and Italy did not wish to be excluded from any air route across the North Atlantic, but had to recognise the advantage enjoyed by the British with possession of Bermuda, and Newfoundland's status within the British Commonwealth. In 1930 the airship R100 made a successful flight to and from Montreal, but in October the R101 crashed in northern France soon after its departure for India and the British government abandoned the airship programme. The Germans, excluded for the foreseeable future from operating either a landplane or a flying-boat across the ocean, continued to operate the *Graf Zeppelin* and steamship-seaplane combinations.

In 1931 Imperial Airways were informed by the Air Ministry that the cruising speed for the proposed flying-boat could not be met, and that the payload which it was hoped to carry was so small that the project was a non-starter. The following year the government cancelled all further work on the Vickers six-engined monoplane. Not only were the authorities responding to the constraints of the financial depression, but Lord Trenchard's influence with the Air Ministry was being exerted, to keep any meagre sums voted by Parliament for the benefit of the RAF.

More hopeful was Sir Alan Cobham. He was aware that experiments to refuel an aircraft in flight had been conducted in the USA. A contender for the world's endurance record had remained aloft for an entire month over California. In 1932 Cobham founded Flight Refuelling Ltd. He was satisfied that an aircraft could take off with a worthwhile payload and cruise over a long range, if a safe method could be devised to transfer fuel to it through a hose from a tanker aircraft. There remained doubts whether this would be acceptable to passengers, and there would also be occasions when bad weather would make it impossible for fuel transfer to be achieved.

Meanwhile the technical manager of Imperial Airways, Major Mayo, developed an idea which had originally been tried by Commander Porte during the First World War. He had taken off in a flying-boat with a single-engined biplane propped above it. Once in the air the biplane was released by a toggle to fly away on its own power. In 1932 Mayo started to design a composite aircraft. Imperial Airways spent three years trying to persuade the Air Ministry to put up half the capital for a prototype before it agreed. Mayo's creation was the four-engined sea-plane *Mercury*, balanced upon the modified Empire flying-boat *Maia*.

The idea of the composite *Mercury/Maia* owed its origin to Commander Porte's own composite built in 1916, using a Bristol Scout. *F. A. A. Museum*

No such stratagems were being considered in the United States where several aircraft manufacturers were developing flying boats, for use over both the Atlantic and Pacific oceans. In 1933 the US Post Office subsidised the American airlines to the tune of £4,750,000. In the summer of that year Lindbergh carried out further surveys for Pan American of both the North and South Atlantic routes. He reported that air services could be run on both of these when a suitable aircraft was ready. Lindbergh's journey followed the purchase by Trippe of the landing concession in Iceland, valid for 75 years. The government of Canada was keen to have an air link with the rest of the British Commonwealth and in 1933 Imperial Airways was granted exclusive rights to operate from Newfoundland for fifteen years. Unknown to the central government, the Newfoundland government had already made a secret agreement with Imperial Airways, allowing the Company the monopoly of landing rights for fifty years.

Not surprisingly Trippe was determined to enjoy the same rights as Imperial Airways, and as the US government was prepared to pay infinitely greater sums to further civil aviation, Pan American was also granted the right to operate through Newfoundland and to make use of the facilities for flying-boats in Bermuda, when these were ready. Trippe then considered another option and made an agreement with Denmark for an air service to Europe through Greenland. In the House of Commons the Under-Secretary of State for Air mentioned that Imperial Airways and Pan American proposed to operate between Bermuda and New York. The Air Estimates for 1934 would show a subsidy of £10,000 towards the service.

In February 1935 Woods Humphery and Mayo met Colonel Shelmerdine and Air Commodore Verney of the Air Ministry technical department to discuss the lack of any British built aircraft capable of flying over the North Atlantic. It had become known that the Sikorsky S42 flying-boat, ordered two years previously by Pan American, was about to be tried out on the sector from San Francisco to Hawaii, and the Martin 130 long range flying-boat was also at an advanced stage of development. Shelmerdine was sure that the performance of these two American aircraft was nothing like as good as was claimed. Woods Humphery had been told by Short's design team that their Empire flying-boat was superior to either of these two contenders. Verney offered the collaboration of the Air Ministry with Imperial Airways on specifications for a flying-boat similar to the Martin 130. Shelmerdine thought that this proposed flying-boat should be used for the carriage of mail.

A few weeks after these conversations the Sikorsky flying-boat flew the 2,400 miles from San Francisco non-stop to Hawaii in 18 hours. Lord Londonderry, Secretary of State for Air, sat up and took notice. He informed Imperial Airways that he urgently awaited their proposals for the North

Atlantic route, so that Great Britain would not find herself unprepared when the Americans were ready. The Company responded with a memorandum highly critical of Air Ministry practices. Pan American's subsidy in 1933 of £1,427,553 was very many times what the British government had granted. The Air Ministry had cancelled the Vickers long range flying-boat. Imperial Airways needed an aircraft with a range of 2,460 miles against a headwind of 40 mph and able to carry a payload of 2,500 lbs. The cruising speed should not be less than 150 mph. A suitable aircraft for the North Atlantic seemed so far away that, in the opinion of Woods Humphery, Pan American Airways should be allowed to open services to the British Isles when they were ready to do so, in order to maintain friendly relations with that company.

Shelmerdine's reaction was to announce that three large flying-boats for the North Atlantic would be built with the Air Ministry contributing to their cost. He did not expect the engines for such a large aircraft to be available in under eight years. Air Commodore Verney was equally pessimistic, referring to the specifications as "1966 and all that". Perhaps recalling the aborted order from Vickers, he suggested that this manufacturer rather than Short's should be given the contract.

The problem was passed to another committee, chaired by the Under-Secretary of State for Air, Sir Philip Sassoon. It was agreed that a large flying-boat should be developed because neither Bermuda nor the Azores had an aerodrome for landplanes. In addition, a prototype landplane should be built at a cost not to exceed £40,000, ready to compete with the Americans when aerodromes were available. The committee was still in session, searching for a way to find enough money, when in November the Martin 130 flew from San Francisco to Hawaii with 1837 lbs of mail, and continued via Midway, Wake and Guam to Manila in the Philippines, thereby inaugurating the first Pacific air mail service. Trippe also anticipated the time when a landplane could be flown on an ocean crossing and placed an order with Douglas. He had been convinced by Lindbergh that landplanes would replace flying-boats, and that aircraft with pressurised cabins would be able to fly above most of the ice-bearing cloud levels. Pressure cabins would be more difficult to include within the hull of a flying-boat.

A month after these American successes representatives of the governments of the United Kingdom, Canada and Eire met in Montreal, and agreed that all aircraft on trans-Atlantic service would land at an Irish airport when travelling east or west across the ocean. The British government would assist in the construction of an airport in Newfoundland and would establish meteorological and wireless facilities there. The Canadians were keen to have Montreal included on the route and to participate in the Empire Air Mail Scheme.

In January 1936 Imperial Airways and Pan American agreed that each

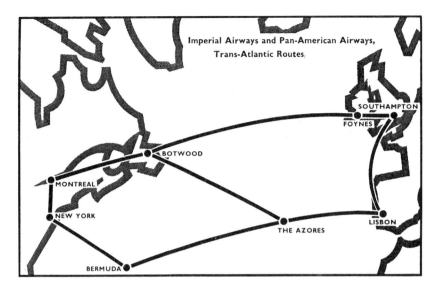

company would operate two round trips each week between New York and England, and that this agreement would last until 1942. In March the Air Estimates included £75,000 for trans-Atlantic bases, £18,000 for the New York–Bermuda service, and £20,000 for the experimental flights to be undertaken by Imperial Airways. Pan American had been keen to initiate trial flights via Bermuda and the Azores since early 1935, but no facilities existed on Bermuda and the Portuguese had not authorised the use of the Azores.

In the autumn of 1936 wireless telegraphy stations were under construction at Rineanna and Foynes, both on the Shannon estuary. Work was also proceeding in Newfoundland, and the early part of 1937 was the target for the commencement of experimental flights. Meanwhile the Air Ministry awarded de Havilland an order for two four-engined landplanes, capable of carrying a payload of 1,000 lbs for 2,500 miles against a headwind of 40 knots.

Imperial Airways proposed to conduct the experimental flights with the Empire flying-boat, which had never been intended for the North Atlantic route. It did not have the range to carry passengers or mail, not even between Foynes and Botwood in Newfoundland. Nevertheless experience could be gained by the crews, flying aircraft stripped of all furnishings to reduce weight, with extra fuel tanks installed to carry over 2,300 gallons of fuel. *Caledonia* and *Cambria* were the two Empire boats chosen for the trial flights.

108

Early in 1937 the Canadian government was at loggerheads with the US Department of Commerce over the choice of a western terminal. The British government agreed with the Canadians that this should be Montreal, and they envisaged an extension to Vancouver and across the Pacific Ocean to Hong Kong. The Americans not surprisingly preferred New York, and there was a lobby which wanted to keep the Pacific Ocean for Pan American to develop without a foreign rival. Approval by Portugal for both Imperial Airways and Pan American to use the Azores, Madeira and Lisbon for their aircraft, pushed this argument into the background. In June 1937 both airlines were almost ready to begin the trials and it was agreed that *Caledonia* and *Clipper III* would take off on the same day, 5th July 1937. Meanwhile the Empire boat *Cavalier* which had been operating in the Mediterranean, arrived in 21 separate crates in Bermuda, was re-assembled and began the long awaited service to New York, in conjunction with Pan American's Sikorsky S42 on 27th May.

It was a disappointment to Air Superintendent Brackley that he was not allowed to command the first westbound flight from Foynes to Botwood. This responsibility fell to Captain Wilcockson who had conducted most of the fuel consumption trials. The latter was seen off by the Taoiseach, Mr. de Valera, on what the Irish would call a 'soft' evening: it was raining. To be sure of avoiding icing conditions in cloud, Wilcockson planned *Caledonia*'s flight at 1,000 feet. He knew that the prevailing wind was from the west and would be stronger if he climbed to a greater altitude.

Captain Wilcockson who commanded the first westbound flight of the Empire flying boat *Caledonia*. July 1937. *Bennett Family.*

109

Travelling in the opposite direction Captain Gray was flying a Sikorsky S42B, which was also stripped of all inessential furnishings and fitted with extra fuel tanks. Neither aircraft carried any commercial load. Gray already had two years experience of long flights over the Pacific Ocean. He had climbed to 10,000 feet and had the advantage of tail winds. For much of the crossing he was able to cruise between layers of cloud, unconcerned therefore with ice-accretion on the wings, propellers and hull. Gray also knew that, within a matter of months, Pan American would obtain the first of six Boeing 314 flying-boats, which would have the power and range to carry passengers and mail across the North Atlantic.

Wilcockson landed at Botwood 15 hours and 26 minutes after departing Foynes. He continued to Montreal where he received a civic welcome, and subsequently flew on to New York. Gray, in *Clipper III*, made the crossing to Ireland in 12 hours 30 minutes and flew on to England. By the end of 1937 Imperial Airways had completed five successful return flights. Pan American completed only four because *Clipper III* was required to replace another Sikorsky which had been damaged in commercial service.

The pilots engaged in these experimental flights speedily became aware of the dangers that the North Atlantic route presented. On one of the flights Wilcockson had been in cloud for nine hours, with a wind of 35 knots reducing the speed of the flying boat over the ocean to under 100 mph. When he made the last westbound crossing in September ice began forming on the wind-screen and the fixed aerial lead-in became a ball of ice. The wireless operator could hear signals only faintly. From the windows of the aircraft the crew saw patches of white ice forming over the wings. The

Pan-Am Sikorski *Clipper*, flown by Captain Gray in July 1937, in Hythe, Hants. *Don Stroud.*

110

airspeed indicator iced up. Every winter the waters around Botwood froze, as did the St. Lawrence river which flows past Montreal. This northern route could never be suitable for year round flying-boat operations.

Despite this Trippe was keen to begin commercial services during 1938 and pressed Woods Humphery to agree. There were other companies in the United States, such as American Export Airlines which had no intention of allowing Pan American to be the sole United States contender on international routes. Nor did these other airlines regard with respect the gentlemen's agreement between Trippe and Woods Humphery, that their two companies would start commercial services on the same day. The United States government declared that each and every service operated to their national territory from Europe would be matched by an American air service to Europe.

Trippe became concerned that his company would be prosecuted for breach of the anti-trust laws if he remained faithful to his agreement with Woods Humphery. Indications that Imperial Airways were not ready brought accusations in the United States that the British were deliberately delaying their service, to keep the *Queen Mary* travelling to and from Southampton full of American tourists.

It was perfectly obvious to Trippe and Woods Humphery that neither the Sikorsky S42B nor the Empire flying-boats would be able to carry a commercial load across the ocean. While Imperial Airways waited to see whether the Boeing 314 aircraft which Trippe had ordered would meet the specifications claimed for it, work was proceeding on three larger flying-boats built by Short's called the 'G' class. Meanwhile the composite aircraft designed by Major Mayo and built by Short's was ready.

On 20th July 1938 Wilcockson took off from the River Shannon in *Maia*. Fixed above him was the four-engined seaplane *Mercury*, flown by Don Bennett with one crew member, Radio Officer Coster. Once airborne, the two pilots released the locking device and Bennett flew *Mercury* on to Montreal with about 1,000 lbs of mail and newsreels. He landed twenty hours later and, after refuelling, flew on to Port Washington, New York to a tremendous reception. A few days later Bennett flew *Mercury* back to England entirely on its own power via Botwood, the Azores and Lisbon.

Imperial Airways had hoped that the Air Ministry would recognise the advantage of such a machine to the Empire Air Mail Scheme, and would support the building of ten or more of them. They were disappointed; but the Secretary of State for Air, Sir Kingsley Wood, was willing to allow Bennett to make an attempt upon the distance record for seaplanes, and on 6th October 1938 *Maia* launched *Mercury* from an airfield near Dundee, and Bennett flew 5,997 miles in just over 42 hours, to land on the Orange River near the Alexander Bay diamond mine settlement in South Africa. Some

Mercury detaching from *Maia* en-route to Montreal in July 1938.

weeks later Bennett flew *Mercury* non-stop from Southampton to Alexandria with Christmas mail.

A few weeks after Bennett's flight to Montreal and New York in *Mercury*, the Germans startled the world with the non-stop flight from Berlin to New York of their Focke Wulf Kondor, in 24 hours 36 minutes. This aircraft had been developed by Luft Hansa for long range flights carrying 26 passengers at a cruising speed of 220 mph. If the second World War had not broken out the following year the Kondor might have been a competitor against the Douglas DC4 four-engined landplane, which was already in production. Imperial Airways considered the acquisition of some Kondors, but were deterred by rising anti-German sentiment, as Hitler acquired the Sudetenland from the Czechs and turned his propaganda campaign against Poland.

Since May 1937 *Cavalier* had been flying between Bermuda and New York covering the 700 miles in a little under six hours. From September this service had been flown twice a week and during the following winter Baltimore replaced New York as the American base. In 1938 Captain G. J. Powell arrived in Bermuda to manage the service and to relieve the other pilot, Captain Alderson, when necessary. During that year various subsidies to help fund the Company's operations from Bermuda reached a total of £54,000, of which the island's authorities contributed £3,320.

In January 1939 *Cavalier* was lost in an accident which also cost the lives of three of those on board. The flying-boat was *en route* to Bermuda in bad weather when carburettor icing caused two engines to fail and the other two began to run down. The pilot, Captain Alderson, landed in the ocean but within fifteen minutes the airliner broke in two in the heavy swell. Two of those on board died of exposure and one drowned, before ten survivors, who had spent ten hours in the sea, were picked up by a tanker, the *Esso Baytown*.

The Inspector of Accidents in his Report upon this disaster could find no excuse for the fact that no satisfactory cure for carburettor icing had been adopted by Imperial Airways. Icing problems of this kind had been reported not only by *Cavalier*'s pilot but by others commanding Empire flying-boats in Europe. The Americans had been able to overcome this problem with their aircraft. The Inspector also criticised the Company's failure to provide inflatable life-rafts, and the absence of any published instructions to passengers in the event of a ditching.

Imperial Airways had intended to replace *Cavalier* with *Champion*, but the months went by without her arrival in Bermuda. Firstly there was the need to solve the carburettor icing problem, and secondly, *Champion* was required to survey the South Atlantic route on her way to the island. The Bermuda Chamber of Commerce not only urged the Company to send another flying-boat, but asked for a larger one because Pan American's Boeing 314 was twice the size of the Empire boats. In November 1938 the

Lufthansa's Fw200 Kondor made a non-stop flight from Berlin to New York in 1938. Imperial Airways considered ordering it but were dissuaded by anti-German sentiment. *Robert Jackson*.

British government announced that the requirements of the RAF took priority over civil aviation, and further trials across the North Atlantic by Imperial Airways would have to be postponed until the following year.

As winter gave way to the early spring of 1939 Trippe took delivery of the first of his company's six Boeing flying-boats. It was reported that these could carry 35 passengers across the Atlantic in considerable comfort at a cruising speed of185 mph. No other civil aircraft could match the performance of the four 1500 horsepower Wright Cyclone engines. On the Boeing the flight engineer could crawl along a catwalk inside the wings to rectify problems with the engines. On one occasion this had been done to deal with a case of carburettor icing.

Imperial Airways had nothing to match the Boeing. The 'G' class flying-boats were far from ready and the costs of these had rocketed, with each engine costing £4,740. It was conceded that Pan American Airways would have to be allowed to begin their commercial service on their own. On 29th March 1939 Captain Gray flew *Yankee Clipper* on a trial run to Southampton via the Azores, Lisbon and Marseille. Within days it became known that de Havilland's landplane project, the Albatross, two of which had been constructed in wood, would be useless on long range routes. The de Havilland Gipsy engines did not deliver sufficient power, and the aircraft could not meet the range specifications. Hopes now rested on the landplane for which Short's had been awarded a contract, the Short 14/38.

Boeing *Yankee Clipper* which won the race to carry fare-paying passengers across the North Atlantic. *Pan American World Airways.*

A Handley Page Harrow refuelling the Short Empire flying boat *Cabot* over Southampton Water. 1939.

By May Imperial Airways had *Cabot, Caribou, Connemara* and *Clyde* ready for North Atlantic trials, and *Cathay* was about to be launched. But the intended date for the Atlantic service was again delayed by problems with the Bristol Perseus engines and failure to overcome carburettor icing.

Pan American could wait no longer. On 17th June 1939 they began a weekly service to Southampton via Newfoundland and Ireland carrying 18 passengers and mail. The single fare was $375. It was proposed to operate via the Azores and Lisbon during the winter months.

Imperial Airways required the co-operation of Sir Alan Cobham's Flight Refuelling Ltd to continue their trials. Two of the Empire boats, *Cabot* and *Caribou*, were stripped of all their furnishings. To compensate for their lack of range they were refuelled by a tanker aircraft based at Shannon. When airborne their normal maximum all-up weight was thereby increased from 43,000 lbs to 55,000 lbs. Two other tankers were stationed at Hatties Camp in Newfoundland, the aerodrome later to become known as Gander. The tankers were old Harrow transports bought by Cobham from the RAF.

The refuelling process was rapid and efficient. It took about twelve minutes to transfer 800 gallons to the flying-boat. This was usually done at 1,000 feet although on one occasion, because of low cloud, the flying-boat was barely clearing the tops of the trees. There were fifteen successful refuelling flights over the Atlantic before they were discontinued. One flight could not be refuelled because bad weather prevented the tanker from finding the flying-boat. The pilots involved in these experiments were Kelly Rogers, Don Bennett and Gordon Store.

The Second World War began before the refuelling trials were completed. Don Bennett was heading west from Shannon when the liner *Athenia* put out an SOS that she had been torpedoed and was sinking. Britain and her Allies ceased to broadcast weather information that could be useful to the Germans, and ships which had always been most co-operative to flight

115

crews, by stating their position and providing wireless bearings on request, resorted to radio silence for their own protection against U-boats.

In his report at the conclusion of these trials Kelly Rogers paid a compliment to the control, wireless and meteorological services which he said "have been in every way satisfactory ... great credit must be given to the Eire government ... all our commanders have expressed their appreciation and complete faith in the officials at both Foynes and Botwood." Kelly Rogers suggested that a vessel should be stationed half way across the North Atlantic, and mentioned that the French government had already positioned a ship with a full complement of meteorologists and the equipment necessary to make a study of the winds and temperature at varying levels. This ship also offered a direction-finding service. After the war weather ships were introduced.

As regards operating height, Kelly Rogers believed that a better aircraft performance than currently available was needed to fly above ice-bearing cloud, and suggested a level of 16,000 to 20,000 feet. In 1939 when he wrote his report there were no aircraft with pressurised cabins. "Speed" he said, "was a vital necessity in the operation of a successful Atlantic service . . . the all-round performance needed would probably be very difficult to obtain in a flying-boat, and it appears that a successful service is much more likely to be operated by landplanes." For the time being however the only alternative to Hatties Camp in bad weather was Moncton, 500 miles further on.

Kelly Rogers was pessimistic about the future prospects of a passenger service dependent upon flight refuelling. He mentioned that on five occasions fuel had leaked into the hull of the flying-boat, and had to be soaked up from the bilges and kept in containers. He drew attention to the risk of fire in such circumstances. He concluded, "Mechanical difficulties have not yet been overcome. Flight refuelling in connection with a passenger service would be out of the question."

In November 1939 the Air Ministry ordered work to be stopped on the Short 14/38 landplane project despite pleas from Imperial Airways, and in spite of the fact that the RAF did not possess any transport aircraft to move troops. This left only the 'G' class in production. The RAF commandeered these when they were ready and *Golden Fleece* was lost when she came down in the English Channel. The cause was icing. *Golden Horn* and *Golden Hind* were eventually returned to the Company, but their range and payload did not remotely match that of the Boeing 314 flying-boat.

Thereafter the airliners built by American manufacturers dominated the North Atlantic skies. Not until the Bristol Britannia turbo-prop aircraft appeared in the 1950s did a British aircraft re-enter the competition, to be followed by the de Havilland Comet IV and Vickers VC10 in the 1960s, and the Anglo-French Concorde supersonic airliner in the 1970s.

116

CHAPTER ELEVEN

Mr. Perkins demands an Inquiry

In June 1937 the death of Sir Eric Geddes, the Chairman of Imperial Airways, happened at a time when a number of the Company's pilots wished to form an association to represent them in negotiations with their employer. Many of these pilots were already members of the Guild of Air Pilots and Air Navigators (GAPAN). This had been founded in 1929 when Sir Sefton Brancker had been installed as the first Master and other members were elected as Wardens of the Court. But GAPAN also included some senior managers of Imperial Airways and their constitution was drawn up to exclude the Court from representing the pilots in the matter of pay and conditions of service.

It was primarily the discontent of the pilots which brought about the attack on Imperial Airways in the House of Commons. This was to result in the formation of a government committee to investigate the charges which M. P.s had heard against the Company. Mention should therefore be made of the general atmosphere of labour relations in Britain during the latter half of the 1930s. The worst years of the financial depression had passed and re-armament was starting to stimulate certain sections of industry. Even so unemployment was widespread, particularly in the areas of declining industries such as textiles, shipbuilding and the coal mines. Jobs were not easy to come by. Employers held the upper hand and many refused to recognise the unions.

Imperial Airways came into the same category as the British Broadcasting Corporation: both were founded after the first World War and, for the great majority of their employees wages and salaries were undoubtedly low, but the prestige attached to these two organisations tended to attract ambitious and enthusiastic recruits who anticipated an exciting future. Imperial Airways engaged well-educated young men as trainees for future postings to stations along the Empire routes. Most of them started work at Croydon where they were paid a token salary of ten shillings a week for their first three years. The unsocial hours of airline duties and the absence of an adequate public transport system obliged them to find lodgings locally and to acquire a reliable bicycle. When they were sent as a traffic assistant to a station overseas their salary rose to £250 a year, but they were forbidden to marry for a further three years. Two of these trainees, Keith Granville and J. Ross Stainton eventually became Chairman of the British Overseas Airways

Corporation.

Imperial Airways had been obliged to recognise some unions affiliated to the Trades Union Congress but had strongly resisted recognition of the union organised by the wireless operators. Unlike the pilots in their smart barathea uniforms this group of airmen were issued with coarse serge clothing, a quarter inch blue ring round the sleeve, but no badge of any kind to indicate that they were part of the flying crew. It was to take eight years before Imperial Airways recognised their union and their status as radio officers. This was brought about by the agreement of British Airways to accord recognition, an action which attracted to that airline some of the Company's men.

Many categories of staff employed by Imperial Airways were liable for duty overseas, but it was not unusual for minimal notice of posting to be given. Wives and families were barely considered and months could pass before they were enabled to join their husbands. This cavalier treatment was suspected of being a deliberate device to save the Company money on the family allowance that was paid to staff based overseas.

The British Airline Pilots' Association (BALPA) came into being following a meeting at the Greyhound Hotel, Croydon in May 1937. A provisional organising committee was formed comprising two pilots from Imperial Airways, two from British Airways, one from Olley Air Services and one from Surrey Flying Services. Lord Chesham consented to becoming president and Lord Amhurst a vice-president. The latter held a commercial pilot's licence and was engaged in aviation in a managerial capacity. BALPA swiftly affiliated to the Trades Union Congress but within a few weeks severed the connection. Probably the longest serving pilots remembered their disillusion with the TUC at the time of their dispute over pay in 1924.

When Geddes had addressed the Company's annual general meeting at the end of 1936 he had reported profits of £140,705. The dividend was to be raised to six per cent and the bonus doubled. He had also said: "For some time the Directors have been quite inadequately remunerated, so I feel sure you will desire to pass a resolution increasing the fees from £6,500 to £12,000 a year." He added: "For the purpose of carrying out our various developments, and particularly the Empire Air Mail Scheme and the Atlantic Services, we shall before our next AGM have to issue fresh capital." Upon his death the succession to the chairmanship of Imperial Airways (and the Dunlop Rubber Company) passed to his deputy Sir George Beharrel. A few days later the latter asked the stockholders to approve the issue of one million pounds in new stock, and declared that the Company would receive three million pounds over the next six years for operating the EAMS.

Throughout the summer of 1937 Imperial Airways had been busily retraining most of their landplane pilots to fly the Empire flying-boats, and

Major Brackley, Air
Superintendent of Imperial
Airways in the 1930s.

also recruiting others. In October a revised pay scale was introduced. Although it turned out that the salary of those currently employed was not to be reduced, new pilots were engaged for a much smaller sum. Very shortly after this announcement six pilots were dismissed. They included the chairman and vice-chairman of BALPA, Lane Burslem and W. Rogers. The latter had been flying for the Company since its inception and earlier still with the Royal Flying Corps and Handley Page Transport. He had refused a request by Woods Humphery that he should resign.

The dismay of the pilots was swiftly communicated to Major Brackley, who in 1924 had been engaged specifically to represent their views to the Board. Since that time long absences on duties overseas and extended proving flights had made nonsense of this function, and he reacted unsympathetically to their approaches. In a letter to his wife he wrote:

I am bitterly disappointed at not being able to see you this weekend, but the pilots are being awkward and a strike is threatened, following a special meeting tomorrow. Although I don't think they will be so ill-advised and foolish as to strike we must be prepared for any eventuality.

119

Two days later he wrote:

The result of last night's meeting gave them little sympathy to strike –they would have been very foolish to do so. It has been staved off, and I think they will see the uselessness of adopting measures only applied by those earning a very small wage . . . they have always had a very fair and square deal and they know it.

Brackley had joined the Guild of Air Pilots at its foundation and in October 1936 had been awarded the Cumberbatch Trophy for Reliability, in recognition of his services to aviation. He made no secret of his distaste at the formation of BALPA, a prejudice undoubtedly shared by Woods Humphery and Burchall.

The Council of BALPA then asked Imperial Airways for recognition of their association and a discussion on the treatment of the dismissed pilots. The Company's reply came in the form of a letter from their solicitors. This declared the dismissals to be justified and denied any necessity for the pilots to belong to a union. This response was criticised by *The Aeroplane*, whose editor C. G. Grey wrote that the interlocking monopoly of Imperial Airways, Railway Air Services, d'Erlangers and the Cowdray–Pearson Group was seen as a danger by their employees. Obviously they would wish to organise unions for their own protection.

The rebuff from Imperial Airways induced BALPA to turn for help to a Member of the House of Commons who sat on the back benches. The House was sparsely attended whenever civil aviation was debated but Robert Perkins, who represented Stroud in the Conservative interest, could be relied upon to be present and to raise questions on inefficiency and malpractice. He held a pilot's licence and flew his own aircraft on business trips to Europe. A few months earlier he had complained in the House of Commons that, on a day of sub-zero temperatures, German, French, Swiss and Dutch airliners had flown in and out of Croydon but, lacking de-icing equipment, no British-built aircraft had left the ground. When Perkins accepted the appointment of vice-president of BALPA he could not have known that his allegations against Imperial Airways would change the course of British civil aviation.

Within days of the approach by the pilots, Perkins had an opportunity to speak on their behalf in Parliament. The date was 28th October 1937 and the occasion was the Debate on the Adjournment:

I want to draw attention to the grave dissatisfaction among pilots employed by Imperial Airways. Their association has been refused the right to collective bargaining. There have been curious dismissals. Captain Wilson, pilot of the "City of Khartoum" which crashed two years ago in the Mediterranean, was dismissed and no reason was given. I feel it was because he opened his mouth too wide when the inquiry took place. Similarly, when the association wrote to Imperial Airways suggesting that the Budapest service be suspended for the winter, as it was not a properly equipped, the two who signed the letter, Captain

Rogers and Captain Lane Burslem, were dismissed. That is victimisation. Imperial Airways have done everything to break the association, and refuse to recognise it or discuss with the pilots at all.

Another grievance is the question of wages. While directors' fees have been increased, they have cut pilots' wages. Then again, the majority of the pilots regard the equipment provided by Imperial Airways as not good enough. With one or two exceptions machines are not equipped to effect blind landings. Few are equipped with de-icing and very few, if any, with a spare wireless set. Finally machines on the London–Paris route are obsolete and certain others on European routes are definitely unsuitable for winter service.

Imperial Airways' service in Europe is the laughing stock of the world. The pilots have tried to negotiate but the door has been slammed in their faces. We have tried to wash our dirty linen in private, but that is no longer possible and we much regret that we have to do it in public. We have no alternative but to ask for an impartial inquiry into the whole position of pilots engaged by Imperial Airways, and in fact into the whole organisation.

The Secretary of State for Air was Lord Swinton, who sat in the House of Lords. Consequently it was left to the Under-Secretary, Lieutenant-Colonel Muirhead, to reply. He said that the government accepted no responsibility for the actions of the Company. The Hambling Report had defined the relationship of the government to Imperial Airways as purely contractual, nothing more. The Secretary of State had asked the government Directors on the Board to investigate the charge of victimisation and they had decided that the dismissals were justified. The government did not consider it necessary to conduct an inquiry.

A few days later Woods Humphery issued a statement denying that membership of BALPA was the reason for the dismissals. The Company was ready to deal with a representative organisation as it did with the radio officers and ground crew. He doubted whether BALPA did represent their pilots; it was not recognised by any other airline. Most of the older pilots preferred direct contact with the Company rather than an association, and there had always been machinery for them to express their views. As regards their aircraft, these were as airworthy as when they were built, easier to fly and more comfortable than any others. The lack of de-icing equipment was due to the absence of any reliable methods of combating ice. The reason why so few airliners had been fitted with SBA receivers for a blind approach was because hardly any airports in England provided that facility. It was not felt necessary to fit aircraft with a spare wireless set when the existing ones were 99 per cent reliable. Finally there had been no reduction in salaries, but new pilots and radio officers had been put on a more realistic pay scale.

Perkins interpreted this statement as an admission that most of his charges were justified. In a letter to *The Times* he pointed out that any de-icing equipment was better than none, and that Cologne, Frankfurt, Prague, Vienna and Budapest were all equipped with SBA.

Robert Perkins, MP who asked for
a public enquiry on the conduct of
Imperial Airways. 1938. *Sir Mark
Norman (stepson of R. Perkins)*

A study of BALPA's records reveals that the letter requesting a suspension of winter services to Budapest by DH86 aircraft, which was signed by the pilots Rogers and Lane Burslem, followed an incident when the latter encountered icing, and lost engine power and radio signals, as his aircraft drifted down into a valley. He was fortunate in that the warmer air below the clouds enabled him to regain power and a safe height. Imperial Airways rostered acting captains on this route but, when it was suspended for the winter, Lane Burslem was dismissed on the grounds that a surplus of pilots had resulted.

Woods Humphery was misinformed about the amount of support BALPA was accorded by the pilots. It was perfectly true that several of the most senior men, who also occupied managerial posts, regarded the leaders of BALPA as troublemakers. A few veteran pilots would not join. But 79 per cent of the Company's pilots had become members, resigning from the Guild of Air Pilots in shoals, and 80 per cent of the pilots employed in regular air services throughout the country had also joined.

Imperial Airways had its defenders. Nigel Tangye, the author and journalist, wrote to *The Times* to support the Company against Perkins' complaints about their old and slow airliners. He attributed this to the priorities of re-armament. Sir Alan Cobham imputed bias to Perkins on the grounds that he was employed by BALPA. But, as the M.P. pointed out in his rejoinder, Cobham had a valuable contract with Imperial Airways for refuelling their aircraft.

This correspondence must have been fresh in the readers' minds when a few days later Sir George Beharrel addressed the AGM of the Company. After announcing an increase in the net profit by £24,000 to £164,735 he declared that, in addition to £12,000 previously allocated for Directors' fees, they were to be rewarded by a further £5,000. A dividend of seven per cent plus a two per cent bonus would be paid to shareholders. The increased dividend was intended to encourage the shareholders to buy the new stock which the Company would soon have to issue. But, not for the first time, its public relations were woefully mismanaged. *The Economist* was one of the publications which criticised this beneficence, complaining that Imperial Airways was hardly behaving in the manner of the public service which it undoubtedly was. *The Aeroplane* noted that government bonds paid $2^{3}/_4$ per cent and, as the Company's dividend was paid from public funds, it should be no higher than that.

On 17th November 1937 the House of Commons listened as Perkins introduced in dramatic terms a motion on civil aviation:

> Like the daughter of Herodias I ask the House today to give me the head of the Secretary of State for Air on a charger. I am asking for nothing less than a public inquiry because I believe that nothing less than a public inquiry will compel the Air Ministry to take some action in the matter. My only regret is that the Secretary of State himself is not a Member of the House of Commons so that he could reply to the charges that I am going to make.

Perkins then embarked upon a series of allegations, beginning with a number against the Air Ministry:

> I am thoroughly dissatisfied with the present position and future prospects of British civil aviation. We know that all is not well. We know that we are behind the Americans and the Germans ... I am inspired simply and solely by a desire to see British civil aviation leading the world, in just the same way as British shipping leads the world now.

Perkins turned to the matter of subsidies which were being paid to Imperial Airways and British Airways for their services to Paris.

> We are faced with the Gilbertian situation of two British airlines, both subsidised by the government, competing with each other, cutting each other's throat, on the same route.

He then went on to explain why British Airways had bought German and American aeroplanes:

> After years of neglect there is no civil airliner of a size suitable to sell in the Empire or in Europe. Tonight, when the night mail air service goes from Croydon to reach Berlin in the

early hours of tomorrow morning, it will be carried by British pilots in a German machine. But that is true not only of this country. It is true of the whole of our Empire.

Perkins informed the House that the Canadians were on the point of signing a contract to buy American machines; South Africa had bought 22 German airliners for their internal services, and Australia was likely to order from the USA.

On the subject of airports, Perkins deplored the fact that the government had for eight years encouraged municipal authorities to build these. They had cost the ratepayers a considerable amount of money, but business was completely stagnant and the Air Ministry had no use for them, preferring to build others for the RAF. Nor did London have an adequate airport:

In the middle of Croydon there is a terrific dip, so deep that an aeroplane can completely disappear from sight . . . it can never be more than a second class airport . . . only worthy of a second class Balkan state.

Perkins said that the only airline which had used Gatwick was British Airways, and after six months they had gone because it was unsuitable. Of this and others he said:

Gatwick and Gravesend are out of the question, Heston is too small.

A further complaint that Perkins directed against the Air Ministry was their acquiescence in the method by which the "octopus" of companies associated with Imperial Airways obstructed the sale of tickets for domestic airlines outside their "ring". The latest victim was North East Airways which served municipal aerodromes:

It is now practically impossible to buy a ticket on that airline. I believe that the policy of the railway companies is to treat civil aviation in the same way as they treated the canals in the early days. Either they want to break them or they buy them out; the idea is to wring their necks.

Perkins went on to express his concern about the amalgamation of the airlines, and his

. . . sympathy for the professional pilots, because it is going to mean in future . . . that every pilot dismissed from one particular airline in this group automatically is dismissed from the whole lot.

In future this amalgamation would have the power to hold a pistol at the head of the government and to say:

Give us a subsidy . . . or we will shut down the whole of British civil aviation . . . I am as good a Conservative as any Hon. Member behind me but I cannot help wondering whether nationalisation of our internal air services would not be preferable to the present position, because as far as I know the whole of British aviation is rapidly coming under the control of two financial houses . . . they are d'Erlangers and Whitehall Securities.

Concluding his allegations against the Air Ministry, Perkins complained

124

that it was

> . . . either unwilling or incapable of doing anything. They have sat back with their arms folded. The fundamental truth is, I believe, that the Air Council who rule the roost have a military outlook, and drag along civil aviation in the same way as a mother drags along an unwanted child. Until we can get civil aviation away from the militaristic outlook of the Air Ministry and hand it where it ought to be, to the Ministry of Transport, we shall never get fair play for civil aviation in this country.

Perkins then turned his attack upon Imperial Airways and began by listing the sources of discontent that had led the pilots to form an association to represent them in negotiations with their employer. Firstly, BALPA opposed the reduction of the salary to be paid to new pilots entering the service. Secondly, it had become a normal occurrence for pilots flying east of Cairo to work continuously for 18 hours on end without sleep.

The pilots wished to discuss with Imperial Airways such matters as the fitting of landing lights on aeroplanes, petrol gauges placed out of sight of the pilot, the malfunction of instruments and engines due to ice accretion:

> These things are not found out by the men who are sitting in offices in London. They are found out only by the men who fly aeroplanes under all conditions . . . certain pilots approached the management about the equipment for the Budapest service and they were promptly dismissed . . . no pilot dare complain because he will be dismissed the next day.

Perkins repeated the charges of victimisation that he had made to the House a few weeks earlier, and then referred to the Guild of Air Pilots and Air Navigators:

> The Guild has done good work in the past, but it is now quickly becoming no more nor less than a city company. In fact it is slowly dying, so engrossed in its own fat that it cannot see out of its own eyes.

After praising the efforts of British Airways to uphold the prestige of the nation's civil aviation, Perkins said that he had been asked whether it was not unpatriotic to criticise Imperial Airways, thereby undermining British prestige. His answer was that:

> Imperial Airways are nothing more nor less than a public utility company. They are heavily subsidised by the taxpayers of this country; they are paying a very large dividend, and they are increasing their directors' fees. Surely Members of this House have not only the right but the duty to subject this company to the most searching criticism and inquiry, in order to see that we get value for money.

Perkins then disputed the officially inspired statements printed in the newspapers, that the pilots of Imperial Airways were paid £1,500 a year. Some captains received as little as £300 annually.

Drawing the Members' attention to the safety record of the Company, he said that the Air Minister had been most reluctant to publish the report of the enquiry into the *City of Khartoum* crash two years earlier. This had revealed that Imperial Airways had been cutting things too fine as far as petrol was concerned:

> Although there were seven major accidents involving loss of life during the last two years, it appears that nine other major accidents occurred, and that only by amazing luck were no lives lost, since in some cases the accidents resulted in the complete destruction of the aircraft.

Anticipating a statement that the European service of Imperial Airways, in comparison with Empire traffic, was a very small part Perkins remarked that six times as many passengers were carried by the European services as on Empire services. Two of those services were suspended during the winter, those to Budapest and to Switzerland. The service to Brussels and Cologne took twice as long as the rival airlines, to Paris twice as long as British Airways.

Perkins deplored the lack of blind landing equipment on the aircraft of Imperial Airways, despite the fact that Heston had been provided with the facility two years earlier, and more recently Croydon also. Referring to equipment to combat airframe icing, he said:

> I wish to be perfectly fair and I know that the Goodrich de-icer is not 100 per cent perfect, but it is fitted to British Airways machines, Royal Dutch Air Line machines and to Swissair machines. It is fitted to ten German machines which regularly fly over the Alps ... its use is, in effect, compulsory in America ... for some reason which I do not understand this de-icer has been more or less barred from Imperial Airways machines.

Perkins admitted that the Dunlop Rubber Company had spent a great deal of money on experiments with their own de-icer, but he thought it wrong that for several years the Goodrich de-icer, however imperfect, had not been adopted. He hoped that the 'Kilfrost' paste which Imperial Airways proposed to smear over the leading edges of the tailplane and wings would prove effective, but he had his doubts. Concluding his speech Perkins said:

> I believe that nothing short of a public inquiry will shake the Air Ministry into a sense of their responsibilities and into a realisation of the present position.

He had been speaking for almost an hour and when he sat down. Lieutenant-Colonel Moore-Brabazon (later Lord Brabazon of Tara) rose to second the Motion. He too was a Conservative M P. and one of the first holders of a pilot's licence. He asked the House not to regard Imperial Airways as a nefarious organisation. Obliged to operate only British aircraft, they had been put in great difficulty by the re-armament programme. The Air Ministry had neglected research. In America civil aircraft manufacturers competed with one another, and the result was the introduction of variable-pitch propellers, flaps and retractable undercarriages. They had transformed civil aviation, adding 100 mph to its speed. Speaking of civil aviation in a general sense Moore-Brabazon declared that "it is far too dangerous.

Accidents are often not reported in the newspapers because they no longer have news value".

Over fifty years later one wonders how many individuals who had just bought an air ticket reacted to that statement. Before the Second World War only a tiny proportion of the public had ever flown, either on business or pleasure. Few could afford to do so and the majority believed flying to be intrinsically dangerous. Insurance companies loaded the policies of those who travelled by air.

For the Labour Party, Mr. Montague said that collective bargaining should be extended to the pilots. He told the House that the air stewards had attempted to form a union under the umbrella of the Transport and General Workers' Union. The reaction of Imperial Airways was to give notice of posting to remote parts of Africa to the leaders. Not one of those who had tried to form the union now remained in the Company. There was an overwhelming case for nationalisation. A Conservative M.P., Colonel Ropner, defended Imperial Airways and said that he would prefer a gentlemen's consultation to a public inquiry. He urged the government to co-operate with the airlines to agree a specification for a civil transport aircraft, and then pay for the prototype.

Lord Brabazon of Tara. Holder of the first civil pilot's licence. *Lord Brabazon of Tara (grandson).*

At the end of the debate the Under-Secretary of State accepted that, in view of the allegations which had been made, a departmental inquiry into charges of inefficiency would be conducted. The Secretary of State (Lord Swinton) would discuss with the two government Directors on the Board of Imperial Airways the system in use for dealing with staff, including the methods by which the grievances of pilots and others were considered. The government would not however look into specific grievances nor dictate to the Company on the recognition of any particular union.

One week later Colonel Muirhead announced that the committee of inquiry would be headed by Lord Cadman of the Anglo-Iranian Oil Company, assisted by two Permanent Secretaries, Sir Warren Fisher of the Treasury and Sir William Brown of the Board of Trade. An objection was raised at once by Clement Atlee, Leader of the Opposition, at the inclusion of the two civil servants. Neville Chamberlain, the Prime Minister, rose to say that he would discuss the composition of the committee with Lord Swinton.

On 30th November 1937, the government announced that the two civil servants would be replaced by Sir Frederick Marquis (later Lord Woolton), Mr. Harrison Hughes of the Suez Canal Company and Mr. Bowen, retired General Secretary of the Union of Post Office Workers. With wide terms of reference the committee began interviewing very many witnesses 'in camera'.

Early in December Imperial Airways submitted a statement of 32 pages with a further 32 pages of appendices. The Board totally rejected many of the charges which had been made against them, including the suggestion that the Company was influenced by the two financial houses, d'Erlangers and Whitehall Securities. As regards the need for a faster aircraft on the European routes, they had been unable to come to an agreement with the Air Ministry which would justify the cost of an aircraft seating sixteen passengers. They had in November 1936 contemplated such an order. Imperial Airways were never shown any of the evidence given 'in camera' against them and so were unable to refute allegations about which they knew nothing. Some members of the Board, Woods Humphery in particular, believed that the issue of BALPA was not the principal concern of the committee at all. Rather an excuse was being sought to bring about a merger with British Airways which was losing a lot of money. They believed that this was the committee's brief.

CHAPTER TWELVE

The Cadman Report

During the summer of 1937 Sir John Reith, Director-General of the British Broadcasting Corporation, let it be known to the government that he wished to move on to another stimulating appointment. Reith took pride in the fact that the BBC was regarded in the highest esteem both nationally and internationally. Later that year however, as criticism of Imperial Airways became widespread, he was disconcerted to hear rumours that he was being talked about as a successor to Sir George Beharrel, as chairman of the Company.

In his autobiography *Into the wind* published in 1949, Reith wrote of his long acquaintance with George Woods Humphery. As young men before the First World War they had both been apprentices in the North British Locomotive Company. Reith's father was priest in charge of the church which Woods Humphery attended and had presided at the latter's wedding ceremony, John Reith acting as best man. They had kept up their acquaintance, meeting for lunch from time to time. According to Reith, Woods Humphery had often expressed great admiration for the BBC when they had discussed the differences in the way their respective organisations were run.

At one such meeting in March 1937 Reith had suggested that the dividend motive in the airline was too predominant, and that Imperial Airways would obtain considerable benefit if they were to employ a top grade Public Relations Officer. He went on to point out the small number of senior managers within the Company, and expressed his opinion that those few were underpaid.

Woods Humphery appeared to agree that the Company's public relations could be better, but reminded Reith that they did not have the financial resources available to the BBC. The Board, he said, would never agree to pay the sort of salary a top grade public relations man would expect. As to dividends, their first duty was to the shareholders. Reith's response to this was that the shareholders would be better served by a properly paid management structure. He believed that the public service motive should take precedence and that the airline would benefit from having a constitution similar to the BBC, which had no shareholders. This view found no support at all from Woods Humphery.

They met again in November 1937. Perkins' recent onslaught on

Sir John Reith (later Lord Reith).
First Chairman of BOAC, 1939 who
replaced Sir George Beharrel and
caused the resignation of Major
Woods Humphery. *Mr. C. Reith (son).*

Imperial Airways in the House of Commons had had an effect, and Reith suspected that rumours of his possible succession to the chairmanship of the Company had reached his old friend. Woods Humphery was insistent that Imperial Airways did not need a full-time Chairman. There was simply not enough for him to do.

Reith was not troubled by this remark because he had neither the desire nor the intention of accepting the appointment, if it were to be offered. In his own words:

> Even if the idea had been attractive otherwise, we were not likely to get on together. His ideas and mine were radically different and I could not have contemplated putting an old friend out of a concern which was largely of his own creation. I reminded him of what I had often urged on him before, for I knew what he did not – that a government committee would soon be appointed to investigate his company and his management.

On 8th February 1938 the Secretary of State received Lord Cadman's Report and early in March it was published, together with the comments of the government. The Report was decisive in its recommendations and robust in its criticism. In the course of the debates in Parliament the relationship of the Air Ministry with Imperial Airways had excited considerable comment, and Lord Cadman recommended that someone outside that Ministry should examine the conduct of their Operational and Intelligence Directorate: that led to the Rae Committee. A further recommendation called for more research by the Air Ministry, whilst development and production of aircraft required civil and military aviation to be more closely related. The Brown Committee was formed to consider this. In a damning comment on the current situation the Report stated: "There is not a medium sized airliner of British construction comparable to leading foreign types".

In respect of air services Lord Cadman wanted Imperial Airways to continue to operate the Empire routes, whilst British Airways should be made responsible for services to all the principal European capitals and also to South America. The London–Paris service should be operated by a third company owned jointly by Imperial Airways and British Airways. Parliament should be asked to provide subsidies for services across the Pacific Ocean and to the West Indies as well as the South American route.

That part of the Report which caused the greatest stir was the committee's criticism of Imperial Airways and in particular Woods Humphery:

> Although the carriage of air passengers in safety and comfort, and the conveyance of mails and freight, have been achieved with considerable efficiency we cannot avoid the conclusion that the management of Imperial Airways has been defective in other respects. In particular, not only has it failed to co-operate fully with the Air Ministry, but it has been intolerant of suggestion and unyielding in negotiation. Internally its attitude in staff matters has left much to be desired. It appears to us that the Managing Director of the Company – presumably with the acquiescence of the Board – has taken a commercial view of his responsibilities that was too narrow, and has failed to give the government departments with which he has been concerned the co-operation we should have expected from a Company heavily subsidised, and having such important international and Imperial contacts.
>
> There should, in our opinion, be an immediate improvement in these respects and this may well involve some change in directing personnel. We further consider that the responsibilities which now confront the Company have increased to the point that they can no longer be borne for practical purposes by a Managing Director. In our view, the Chairman of the Company should be in a position to give his whole time to the direction of the business and should do so. We think the Chairman should personally control the management of the Company and he should be aided by the services of one or more other whole-time Directors.

The Cadman Report inevitably brought a wide variety of reactions. C. G. Grey, the best-informed aeronautics correspondent, had attributed the cause of Imperial Airways' troubles to "the shadow of Sir Eric Geddes", even before the publication of the Cadman Report:

> His principle, which may have brought him to his outstanding success in the material things of life, was to build up a system of almost inhuman mechanical efficiency. The one thing that is needed when you are dealing with such temperamental people as the pilots and the crews of aeroplanes is the intimate personal touch. They need a dictator, but he

must be . . . a personal leader which Sir Eric Geddes was not. To the employees of Imperial Airways he was a machine to be feared rather than a human being to be loved and followed. Even many of those who admired him, because of the kind of outward efficiency which he forced on to other people, could never bring themselves to regard him as a human being.

C. G. Grey declared that the Cadman Report was "the most sensible document that has yet been issued on civil aviation". But he remained steadfast in support of the beleaguered Woods Humphery:

Precisely what is the difference in function between a full-time Chairman and Managing Director I cannot imagine. My idea of a Chairman is a Director who comes to the Board meeting, swings a bit more influence than the other Directors do, and goes away. But a full-time Chairman must be much the same thing as a Managing Director. So the best thing to do would be to make George Woods Humphery full-time Chairman and let the Board then find a couple of other full-time Directors. The great thing is to find some way of taking some of the work off his shoulders. But, quite definitely, his resignation would be the very worst thing that could happen for Imperial Airways Ltd and for British civil aviation.

Before the end of March 1938 the Company sent a letter to all their shareholders to express their shock at the Report's strictures upon their Managing Director:

Mr. Woods Humphery has been condemned by the committee without having an opportunity of saying a word in his defence on the matters in question.

In its own defence the Board told the shareholders:

There is the bald statement that your Company is operating in Europe with obsolete aircraft. This is entirely beyond the control of your Company, and it is to be regretted that no explanation as to why obsolete aircraft are running on certain European services is given, although evidence as to the causes was tendered fully before the Committee, and no reference whatever is made to the fact that for over 90 per cent of the routes operated by your Company, the services are maintained almost entirely by the most up-to-date flying-boats in the world . . . there was no lack of foresight on the part of the Board in this matter, for in 1934 an order was placed with an important British company for twelve large landplanes, comparable in every way with the flying-boats which are now so successfully operating the Empire routes. The first of these landplanes was due for delivery in September 1936 but in spite of the efforts made by the Air Ministry and your Company not one has yet been delivered.

The Board rejected allegations that it had been "intolerant of suggestion and unyielding in negotiation." It documented examples of the way the Air Ministry had delayed paying agreed subsidies, failed to answer letters and conclude contracts. Negotiations for the Empire Air Mail Scheme had been extended over four years, and a decision over the use of Langstone Harbour as the Empire flying-boat base did not come from the Air Ministry until two years had passed.

The spokesman for the Captain's Committee of Imperial Airways wrote to Sir George Beharrel:

My committee and the captains express their unanimous satisfaction in the management of Imperial Airways, with particular reference to the Managing Director, who has been subject, in their opinion, to unwarranted criticism by the committee of inquiry – Indeed they are glad if there have been times when the management has been intolerant of suggestion and unyielding in negotiations with the Air Ministry.

In his charges against Imperial Airways that had led to the committee of inquiry, Robert Perkins had stressed the discontent of the Company's pilots. The Cadman Report contained the following statements:

In our view personal contact must now be supplemented by collective representation of employees. The desire for such a change has been expressed to us by the representatives of the pilots. Imperial Airways . . . has stated that it has no objection to collective bargaining . . . it is essential however that any organisation formed with these objects should be in a position to negotiate authoritatively on behalf of a substantial proportion of the class it claims to represent.

Imperial Airways chose to believe that BALPA was not supported by a sufficient number of their pilots and refused to recognise it. Meanwhile the government tried to find a willing candidate to take over, full-time, the position of Chairman of Imperial Airways. One of the governments nominated Directors, Sir John Salmond, was approached and declined the post as did Sir John Anderson.

In the House of Commons the Prime Minister, Neville Chamberlain, asked M.P.s to accept the Cadman Report. The large Conservative Party

Prime Minister Chamberlain's return from Munich in September 1938 on a Lockheed 14 Super Electra of British Airways.

majority ensured that this appeal would be successful. The government proposed to increase the annual subsidy to £3,000,000 and to respect also Lord Cadman's recommendations. At the end of the debate the Under-Secretary of State for Air said that the matter of the pilots who had been dismissed was within the Company's competence to settle, and was not the business of the government. Both Imperial Airways and British Airways fell into line, accepted the Cadman Report, and agreed to the appointment of a full-time Chairman.

When Sir John Reith was made aware of the criticism of Woods Humphery and the Board of Imperial Airways he was, in his own words, "less inclined than ever to go there". He assumed that Woods Humphery would either offer his resignation or be asked to do so. He had misjudged his old friend. The Managing Director, with the support of Beharrel and at least some members of the Board, did not resign, declaring that his sense of responsibility to the shareholders would not permit him to abandon them.

In May 1938, the Prime Minister responded to the demand that the Secretary of State for Air should be available for questioning in the House of Commons. The Air Ministry had become the second largest spending department as a consequence of re-armament. Sir Kingsley Wood was appointed in place of Lord Swinton, and Harold Balfour resigned as a Director of British Airways to take office as Under-Secretary of State for Air.

In June, Woods Humphery visited Reith again, convinced that the latter would be offered the post of Chairman. Once again he emphasized the prime responsibility to the shareholders, to them and no one else.

Throughout the summer of 1938 Adolf Hitler had been increasing his pressure on Czechoslovakia to cede the Sudetenland to Germany The government of France was terrified that the Czechs would resist and demand that the French honour their pledge to come to their defence. Daladier, the Prime Minister of France, was looking to Chamberlain to help him persuade the Czechs to accept some formula, any formula, to avoid war. The problems of Imperial Airways cannot have occupied many of Chamberlain's waking hours.

Reith had been to see Leslie Hore-Belisha, Minister for War, to ask for some responsible job at the War Office. The latter had warmly welcomed his offer, but Sir Warren Fisher (Permanent Secretary to the Treasury) prevented any such appointment being confirmed. On 3rd June 1938, Reith received a telephone call from Sir Horace Wilson, who had become Chamberlain's closest adviser, although his official role related to labour affairs. Wilson told Reith to call upon Sir George Beharrel and to accept the chairmanship of Imperial Airways. He added that the resignation of Woods Humphery should be insisted upon.

In his autobiography Reith declared that he had no wish to accept the appointment. He had an interview with Chamberlain lasting only four minutes, just time enough to make it clear that he had no desire to take on the job. He was only prepared to accept if that was the wish of the government. Being assured that this was so he went to meet Sir George Beharrel at the Athenaeum Club. The latter told him that his fellow Directors might not be as welcoming as the government to Reith's appointment. It was most regrettable that the post of Chairman would be full-time. There would not be enough for him to do. Reith put his discussion with Beharrel on record:

> I asked why he and his Board had accepted the Cadman recommendation about a full-time Chairman, but there was no answer. I enquired about Woods Humphery and was told that he was certainly staying on; the Board could not contemplate him leaving, I asked Beharrel to face up to the situation. The post I was going to was that of full-time executive Chairman, that is Chairman and Managing Director: but there was a Managing Director already who, according to him, must stay. He replied that there would need to be some allocation of work between Woods Humphery and me. I told him there could be no restriction of my authority if I went there. I would not be content with what any official decided to refer to me. How would he feel himself in such a position? He said he certainly would not like it; most invidious.
>
> When he enquired what my answer was I said he could provide the answer himself. The whole situation was distasteful and embarrassing to me, especially as I had known Woods Humphery for so long. When he asked what salary would be required I said I was getting £10,000 from the BBC, and would not require any increase on that. He 'thought that they might be able to go to that figure but no more'. He did not mention that Woods Humphery was already getting £7,000 plus £1,500 expense allowance, and that if he had still been there in September he would have had £10,000 plus £1,500.

Reith left the meeting feeling too miserable to attend a Toscanini concert in Queen's Hall that night.

His dejection was not lifted when he paid his first visit to the Head Office of Imperial Airways:

> I was brought to the door of an old furniture depository behind Victoria Station. It was Imperial Airways – a plate on the wall said so. Inside were some counters, luggage on the floor, a few people standing about – a booking office evidently. I enquired of a young man behind one of the counters where the Head Office was. He pointed to a dark and narrow staircase – up there, he said. The Managing Director's office, second floor he thought. Having ascended thither I went along a passage, also dark and narrow, between wooden partitions, peering at the doors and wondering which to try first. Here it was – a bit of paper with "Managing Director" written upon it. From Broadcasting House to this.
>
> And the first decision demanded of me was an indication of what had happened to me otherwise. Would I approve the expenditure of £238 on passengers' lavatories at Croydon? I enquired politely if such a matter and such a sum had in the past required the approval of the Managing Director personally. Yes indeed. It seemed I was to work in very low gear. I doubted my capacity . . . If ever I had visited Woods Humphery in his quarters I should have drawn conclusions which might have made me decline even to consider Imperial Airways, when the idea was first mooted, instead of letting things go by default.

As news of the appointment of the new Chairman spread, Sir Montagu

Norman, Governor of the Bank of England, sent a message of approval to Reith and urged him to insist upon Woods Humphery's resignation. So too, according to Reith, did Sir Warren Fisher, the Head of the Civil Service. Norman also advised the transformation of Imperial Airways into a public corporation, an action to which Reith had already committed himself.

When Sir Kingsley Wood told the House of Commons that the Board of Imperial Airways had decided to appoint Reith as the new Chairman, and that Sir James Price had been asked to join the Board in order to investigate accusations of management wrongdoing, Woods Humphery offered his resignation. He was asked and agreed to stay on long enough for Reith to settle in. His decision to resign was inevitable once it had been made clear to him that he could remain only as General Manager without a seat on the Board. Even that might have been practicable if Reith had been allowed, as he wished, to retain the post of Director-General of the BBC. That was denied to him.

The staff of Imperial Airways closed ranks and expressed their dismay at the manner in which Woods Humphery had been made the scapegoat for a prolonged series of misfortunes. The 3,600 staff spread over 111 stations sent cables deploring the resignation. On 24th June 1938 a testimonial dinner concluded with a presentation of a silver salver from members of staff to Woods Humphery. Captain Wilcockson spoke of "a dirty political move" Colonel Burchall said: "We have intense indignation at the campaign of calumny waged against the Company and our chief for whom we have respectful affection and loyalty . . . Under his guidance the finest commercial aviation service in the world has been produced." From the other side of the world Hudson Fysh cabled: "Resignation Woods Humphery at this stage inauguration Empire Air Services fantastic and unacceptable. Do hope British stability will prevail."

Assistant General Manager Dismore wrote to the Prime Minister, referring him to the Board's 'Comments on the Cadman Report' and demanded that the Company's accusers should plan ten years ahead. *The Observer* took his point: "Mr. Woods Humphery was charged with making commercial aviation pay and he came nearer to succeeding than anyone else in the world. The new executive Chairman, Sir John Reith, is charged with making British aviation technically advanced, no matter whether it pays or not " C. G. Grey suspected that the government wanted Woods Humphery removed because "he was the effigy and victim of Sir Eric Geddes".

It is not an easy task, so many years after these events, to understand why the government was so determined to force Woods Humphery out of the organisation to which he had devoted so many hours of each day since 1924. The constitution of Imperial Airways was decided upon by the government-appointed Hambling Committee. The Managing Director had to work

Left to right. Major Mayo, Major Woods Humphery and Lieutenant Colonel Burchall on the occasion of Woods Humphery's departure from Imperial Airways following the appointment of Sir John Reith as Chairman. 1939.

within a framework of shareholders expecting a return on their investment, and the government expecting an efficient air service at the lowest possible cost. If the shareholders could be kept contented, then the taxpayers could complain that the dividends came from their pockets. Yet, even allowing for inflation, the annual subsidies did not compare in any way with the million pounds a day aid to the Coal Board by governments in recent times.

No one could charge Woods Humphery with inefficiency. He could be abrasive, but he had to deal with individuals outside the Company who did not share his priorities. Within Imperial Airways those who worked closest to him were his loyal and devoted followers. They included the most senior captains. He was a very private person who avoided close friendships with his staff. It was a rule that no two members of the same family were employed. Undeniably he had an aversion to unions, but that was not unusual among employers in the 1930s when many preferred to pay wages above the union negotiated rates in order to avoid the continuous distraction of pay negotiations which later bedevilled British industry.

The Company's dependence upon the Air Ministry was a major cause of its problems. The Air Ministry had ample evidence of the usefulness of transport aircraft for military purposes, but in 1938 the small number of such machines which the RAF possessed owed their origins to a twenty year

Maia and *Queen Mary* at Southampton in June 1938.

old bomber. The Handley Page Harrow which had gone into production in 1935 was obsolete as soon as it appeared, not remotely comparable to American aircraft of the period. One reason was the reluctance of Lord Trenchard to devote Air Ministry funds to aircraft the Army might require for troop movements. Civil servants in the Air Ministry were accountable to the Treasury and to their political masters, and without experience of commercial considerations.

It was hardly surprising therefore that no aircraft was developed specifically for commercial operation until the re-armament programme had also got under way. Even then, preoccupation with short stage operation on the Empire routes left Imperial Airways without an airliner suitable for passenger carriage across the North Atlantic Ocean. After the six-engined Vickers flying-boat had been cancelled it had taken years to persuade the Air Ministry to build Major Mayo's composite, only for its production as an air mail carrier to be abandoned after *Mercury* had been built and flown.

In this situation the small number of the Company's staff, mostly underpaid, were driven hard while the Board of Imperial Airways waged a constant battle with the Air Ministry, itself under-funded, to obtain suitable aircraft to compete against the heavily subsidised foreign airlines. Undoubtedly there were those in government who feared that if Woods Humphery, who knew more than anybody about those who were obstructing the Company's progress, was promoted to Chairman he would become even more of a thorn in their flesh than when Geddes had been alive.

138

CHAPTER THIRTEEN

The Short Reign of Sir John Reith

One beneficial result of the Cadman Committee's strictures over the aircraft procurement policy was the issue by the Air Ministry of two specifications for four-engined airliners. In the past British manufacturers had delivered aircraft quite rapidly once their tenders had been accepted, and the in-service date for both these new machines was set for 1940. The order for a long-range airliner was given to Short's. The target speed in the cruise was designed to be 275 mph but if a pressurised cabin was incorporated then a speed of 330 mph at 25,000 feet was specified. The aircraft was to have a range of 3,000 miles over which a payload of 7,500 lbs could be carried. The contract for an airliner for the European routes was awarded to Fairey who offered a design known only as FC1. In addition to the two prototypes, twelve production machines were ordered.

When Reith took over as Chairman in June 1938 the grievances of the pilots remained unresolved. In the House of Commons M.P.s continued to ask questions, particularly about the treatment of Captain Rogers. Finally, Imperial Airways agreed to pay him a retainer and to re-instate him when the new company, recommended by Cadman, began operations between London and Paris. This company was never established and Rogers became private pilot to Lord Forbes. Lane-Burslem was engaged by de Havilland. A ballot conducted by the pilots showed that about 65 per cent of the captains and almost 90 per cent of the first officers wished BALPA to be their representative agency in negotiations over pay and conditions. The Company was in no hurry to accept the results of this ballot and a captain's committee unsuccessfully tried to obtain an improvement in the salary scale, or an allowance for the increased cost of living.

The charges made against the late Sir Eric Geddes, Woods Humphery and Imperial Airways had to be investigated. In the main these were all repudiated or adequate explanations accepted. A new Director, Sir James Price, was given the task of establishing an improved pay scale, whilst Reith himself held conferences with the pilots to hear their complaints.

As for the existing structure of Imperial Airways, the new Chairman declared himself greatly disappointed. He had been led to believe that it had an organisation which he would understand, devolutions of responsibility, a clear administrative system. He wrote:

> What I found was Woods Humphery personally handled an inordinate amount of detail, so that to a great extent the whole business had depended upon him. With his departure no one in Imperial Airways knew where they were.

Reith remained determined to transform Imperial Airways into a public service corporation like the BBC. In his own words: "I will not let the shareholders down; I will let them out, with good terms for their going." He told Beharrel and the Board that having recently paid the shareholders a nine per cent dividend the public service motive would in future take precedence.

In a departure from Lord Cadman's recommendations Reith's own plans envisaged an amalgamation of Imperial Airways with British Airways, to make one air corporation rather than the retention of two companies, each with a full-time Chairman. He discussed this with officials of the Treasury and Air Ministry in July, and in the following month the Hon. Clive Pearson, Chairman of British Airways, agreed to a scheme for amalgamation. In October the Board of Imperial Airways unanimously accepted these proposals.

When the Cabinet came to consider the matter Sir Kingsley Wood expressed reservations about the proposed emergence of one large nationalised air corporation. Obviously he felt uncomfortable about the abandonment of traditional free market Conservative principles. News of the Minister's objections reached Reith, who reminded the government that he had accepted the chairmanship of Imperial Airways on the understanding that it would be managed as a public service corporation, in other words, a nationalised undertaking. Finally, under pressure from colleagues in the Cabinet, Wood went along with Reith's proposals.

On 11th November 1938 the Minister gave details of the agreement to the House of Commons, telling M.P.s that the shareholders of each company would be asked to approve a fair and reasonable price for their holdings. The new public corporation would obtain funds to purchase the two companies and, for future expenditure, by the issue of fixed interest stock, guaranteed by the government. *The Times* and *The Economist* approved of the outcome; *Aeroplane* took rather longer to express satisfaction that the government had at last accepted responsibility for the air services to the Empire.

When the time came for Reith to address the shareholders he explained why a non-commercial constitution was necessary for civil aviation. He reminded them that the Secretary of State had told Parliament that a considerably greater capital injection was needed. This would have meant a great increase in ordinary share capital, or the creation of shares ranking before those currently held. There would therefore be a seven per cent dividend, and the shareholders and Board would meet again before the price to be paid for their shares was agreed.

Thanks to a helpful attitude by the Treasury this did not prove to be a serious problem, and eventually they received almost four shillings per share more than the current market price. It was less easy to fix a price which the government would pay for the two companies. In the end it paid £2,659,086 for Imperial Airways and £262,500 for British Airways. The latter had lost money in both 1936 and 1937, but thereafter had begun to make profits. With all the financial details agreed the way was cleared for the introduction of the appropriate Bill to Parliament in June 1939.

Meanwhile Reith had chosen the name of the merged airline – British Overseas Airways Corporation. It would take over from the two companies on 1st April 1940. C. G. Grey, writing in *The Aeroplane* on the demise of Imperial Airways thought this title appropriate: "I give full credit to that great Scotsman, John Reith, for the sardonic humour which chose the word BOA for the corporation which swallowed it up." Along with everything else BOAC took over from Imperial Airways the speedbird symbol, introduced in 1938.

Reorganisation of the management, as the amalgamation of the two companies was put into effect, brought a severe disappointment for Major Brackley. The post of Air Superintendent was abolished and he was transferred to serve under Campbell Orde of British Airways, who was designated operations manager. *The Times* reported: "This change releases Major Brackley for other and special duties for which his experience fits him well. In future he will undertake, among other things, the survey and development of new routes." With the German annexation of Czechoslovakia bringing war in Europe that much closer Brackley was sent to carry out a survey along the West African coast, in anticipation of a future diversion from the existing routes, should hostilities disrupt the existing links to the Empire.

Woods Humphery, loyally fulfilling his promise to stay on until Reith had completed the re-organisation, was returning from New York on the *Aquitania,* having visited Juan Trippe for discussions on developments for a North Atlantic service. In a letter to his successor he expressed his disappointment that the three G-class Empire flying-boats, which should have been in service on that route in 1938, were still in the development stage, whilst Pan American's *Yankee Clipper* had already made a proving flight to England. He reminded Reith of his own efforts to obtain a genuine trans-Atlantic airliner and emphasized the enormous amount of capital expenditure that would be needed to establish the route. "As you have now seen," he wrote "we have been very badly treated by the Air Ministry for over two years, as regards inaction and indecision." He told Reith that the range limitation of the Company's Empire flying-boats would require them to touch down in the Tagus river at Lisbon and again at the Azores. Pan American Airways had already provided themselves with jetties, mooring buoys and launches to handle their Boeings. Woods Humphery enquired whether he should ask Trippe for permission to share these facilities. He concluded with the words: "Perhaps these few idle thoughts of an idle man may be of use to you."

Reith's reply, when it came, was singularly terse. Acknowledging "your longhand note", he went on to say: "Let us adopt the memorandum form in writing to one another. It is simpler and quicker – if you agree." He admitted that he had considerable power but claimed something else was required: "a Secretary of State who is prepared to overrule the military side and guarantee that civil aviation has its proper place". Woods Humphery's response to this was suitably brief: " I am much obliged to you for your memorandum which is most informative and interesting."

At the end of July 1939 the first of the G-class flying-boats, *Golden Hind*, made its initial flight and very shortly afterwards was handed over to Imperial Airways for the conduct of proving trials. But de Havilland's Albatross, the four-engined airliner of wooden construction which had originally been considered a possible contender on the Atlantic route, was having its problems. The best one could say in its favour was that it could be depended upon to compete in speed with the Company's rivals on the European routes,

As Hitler fulminated against Poland and demanded the return of Danzig to the German Reich, the Hon. Clive Pearson, who had been Chairman of British Airways, became a Director of the combined airlines and Deputy Chairman to Reith. Also appointed as Directors were Mr. Harold

Golden Hind one of three Short Bros G-class flying boats.

Captains Don Bennett and Tony Loraine at Boucherville, Canada a few days before the outbreak of war on one of the last flying boat trials over the Atlantic. *Cabot* in background was destroyed during the Norwegian campaign a few months later.

Brown and the Hon. Leslie Runciman, an amateur pilot who held director-ships in Lloyds and the London and North Eastern Railway.

At midnight on 30th August 1939, on the eve of Germany's assault on Poland, the entire eastern half of England became a prohibited area for civil aircraft. Imperial Airways and other airlines were ordered to evacuate Croydon. Every Company aircraft, except for three which were unservice-able, was flown to Whitchurch near Bristol and the Grand Spa Hotel in the city was commandeered for use by the Company as their headquarters. Ten days later the HP42s and DH89s were transferred to Exeter Airport. Imperial Airways was ordered to move the flying-boats from their base at South-ampton Water to Poole harbour, where sailing club premises were taken over and moorings put down.

The *Golden Hind* which was being used for crew training, was comman-deered by the RAF for maritime reconnaissance. Both *Caribou* and *Cabot*, which had been engaged on the flight refuelled Atlantic trials, were also taken over for use in anti-submarine patrols. The Empire flying-boats still based in England were stripped of sound proofing and much of their furnishings to provide increased cargo space. To increase the take-off weight by 4,000 lbs more powerful Pegasus engines were fitted. The Company was authorised to run a daily service between Heston and le Bourget with passengers on official business. A route from Perth to Oslo and Stockholm was also operated, using British Airways Junkers Ju52s and a Lockheed 14.

Reservists were called up and a number of airline pilots and members of other aircrew categories, who were in the RAF Reserve, departed to various depots and air stations. Brackley was 45 years old, but when he returned from the African survey he applied to the RAF and was sent to Coastal Command with the rank of squadron leader.

Despite protests from Reith the Air Ministry cancelled all further work on the Short and Fairey landplanes. From the British embassy in Washington and in the pages of the *New York Times* came urgent appeals for some sort of air service to be continued across the Atlantic in the interests of propaganda for Great Britain. Reith kept up the pressure, informing Woods Humphery that

> we are forced back to the G-class since we really have nothing else . . . we are fully alive to the importance of doing something on the Atlantic, and are doing what we can to meet the deplorable state of affairs, but without much success so far.

There was great disappointment that the Under-Secretary of State for Air, Harold Balfour, who had until so recently sat on the Board of British Airways, had done little or nothing to prevent the erosion of the Company's resources. It is only fair to say that Balfour, upon his appointment in the summer of 1938, had been horrified to discover how unprepared were Britain's air defences. The 17 new stations which were intended to provide radar coverage from Newcastle to Land's End were not completed. Of the regular fighter squadrons 19 were still equipped with biplanes and only two with Hawker Hurricanes. There were insufficient Supermarine Spitfires to equip a single squadron. It was hardly surprising therefore that he did not accord a high priority to submissions on behalf of the airline.

The autumn of 1939 turned into a bitterly cold winter and in the early spring of 1940 there had still been no fighting of any consequence on the western front. Woods Humphery had left for the United States to join another former director of Imperial Airways, Hubert Scott-Paine, in a successful motor-boat venture. Major McCrindle, Reith's deputy, anticipated that "the public may wake up one day to find to their surprise and consternation that the airline has gone out of business."

When Balfour was asked in Parliament whether he could make any statement with reference to the Atlantic services during 1940 he replied:

> It had been hoped to reopen this important service during the present year. Recent developments here however have made it necessary to direct the aircraft intended for use on this service to certain defence purposes for which they are particularly suited. I regret that at the present defence needs must claim priority.

144

Major Brackley (right)
visits an African station
with Sir Hubert Walker
(left), Director of Civil
Aviation, Nigeria.

The Air Ministry had at last discovered how blind had been those mandarins who had overlooked the need to provide any aircraft for moving military personnel and their equipment in the event of war. As long as Imperial Airways was dependent upon this and other government departments for a large measure of funding and aircraft procurement the airline could not possibly develop as it should. On 1st April 1940 BOAC was officially born, but in the most unfortunate and depressing circumstances. Imperial Airways ceased to be, but the Company's best asset had been a few thousand loyal and enthusiastic staff. In 1994 over 100 of those former employees met for one of their regular reunions to share their memories and to recall the pioneering days of British civil aviation.

Epilogue

Woods Humphery was able to make one further important contribution to British aviation. He urged the British Purchasing Commission in the United States to arrange for the Lockheed Hudson twin-engined aircraft to be flown across the North Atlantic from Canada, thus speeding up delivery and avoiding their loss in those ships which succumbed to torpedo attacks by U-boats. Although it was anticipated that the severe winter weather and engine failures would take their toll this idea was accepted. Lord Beaverbrook, Minister for Aircraft Production, asked Sir Edward Beatty, Chairman of Canadian Pacific Railways, to organise the air ferry service, and Woods Humphery was invited to act as deputy to the Chairman.

Colonel Burchall had been appointed Deputy Director-General under Reith's reorganisation, but he resigned to take up an appointment as General Manager on the North Atlantic Ferry Organisation, delighted to assist his former Chief.

Major Brackley had risen to the rank of Air Commodore as Senior Air Staff Officer in Transport Command when the war ended in 1945. He rejoined BOAC as a member of the Chairman's staff until he was invited to be Chief Executive on the Board of British South American Airways in March 1948. In November of that year he lost his life in a drowning accident in Rio de Janeiro. Sidney Dismore, the long serving Company Secretary, together with D. H. Handover, the traffic manager and R. Waugh, the sales manager, resigned after less than three years service with BOAC.

Sir John Reith was awarded a peerage and left BOAC in May 1940 to become Minister of Information. The Hon. Clive Pearson, Reith's successor as Chairman of BOAC resigned in 1943, when all BOAC operations were made subservient to RAF Transport Command. So too did fellow Directors Runciman, Brown and I. C. Geddes.

The policy of building mainly bombers and fighters left BOAC at a great disadvantage when peace returned. They were obliged to carry passengers in converted bombers as in 1919. The government had no option but to allow BOAC to buy a succession of American built aircraft to compete on the North Atlantic service. In 1948 the first British-built pressurised airliners, the Avro Tudor and the Handley Page Hermes, were not suitable for this route. It was 1958 before a British-built airliner finally inaugurated a regular service to New York–the turbo-prop Bristol Britannia. This was followed by the pure jet de Havilland Comet IV in October of that year. Later the Vickers VC10 and the supersonic Concorde made a good impression, but in terms of numbers Boeing aircraft still dominate the North Atlantic skies.

Airliners operated by Imperial Airways, 1919–40

in chronological order

type and registration	name	taken into service	taken out of service
de Havilland DH34			
G-EBBR	*City of Glasgow*	from Instone 1924	destroyed, Ostend 1924
G-EBBT	*City of New York*	from Instone 1924	scrapped 1926
G-EBBV	*City of Washington*	from Instone 1924	scrapped 1926
G-EBBW	*City of Chicago*	from Instone 1924	scrapped 1926
G-EBBX	(no name)	from Daimler 1924	crashed, Purley 1924
G-EBBY	(no name)	from Daimler 1924	scrapped 1926
de Havilland DH50			
G-EBFO	(no name)	1924, used on charter	returned to de Havilland, 1924
G-EBFP	no name	from DH 1924	sold 1932
G-EBKZ	(no name)	1925, used on charter	crashed, Plymouth 1928
Handley Page			
W8B G-EBBG	*Princess Mary*	from Handley Page Transport	crashed, Abbeville 1928
W8B G-EBBH	*Prince George*	from Handley Page Transport	scrapped 1931
W8B G-EBBI	*Prince Henry*	from Handley Page Transport	scrapped 1932
W8F G-EBIX	(no name)	1924	crashed, Boulogne 1930
W9 G-EBLE	*City of New York*	1926	sold 1929
W10 GEBMM	*City of Melbourne*	1926	sold 1933
W10 G-EBMR	*City of Pretoria*	1926	sold 1933
W10 G-EBMS	*City of London*	1926	crashed, English Channel 1926
W10 G-EBMT	*City of Ottawa*	1926	crashed, English Channel 1929
Supermarine Sea Eagle			
G-EBGR	(no name)	from British Marine 1924	withdrawn 1929, written off 1930
G-EBGS	(no name)	from British Marine 1924	sunk, Guernsey 1927
Armstrong Whitworth Argosy			
G-EBLF	*City of Birmingham*	1926	broken up 1934
G-EBLO	*City of Glasgow*	1926	crashed, Assuan 1931
G-EBOZ	*City of Wellington*	1927	withdrawn 1934

G-AACH	*City of Edinburgh*	1928	crashed, Croydon 1931
G-AACJ	*City of Manchester*	1928	sold 1935
G-AACI	*City of Liverpool*	1928	crashed, Dixmude 1933
G-AAEJ	*City of Coventry*	1928	written off 1934

de Havilland DH66 Hercules

G-EBMW	*City of Cairo*	1926	crashed, Timor 1931
G-EBMX	*City of Delhi*	1926	sold 1935
G-EBMY	*City of Baghdad*	1926	scrapped 1935
G-EBMZ	*City of Jerusalem*	1927	crashed, Jask 1929
G-EBNA	*City of Teheran*	1927	wrecked in gale, Gaza 1930
G-AAJH	*City of Basra*	1929	sold 1934
G-AARY	*City of Karachi*	1930	scrapped 1935
G-ABCP	*City of Jodhpur*	1929	crashed, South Rhodesia 1935
G-ABMT	*City of Cape Town*	1929	sold 1934

Short S8 Calcutta

G-EBVG	*City of Alexandria*	1928	capsized, salvaged, sold 1936
G-EBVH	*City of Athens*	1928	sold 1937
G-AADN	*City of Rome*	1928	sank, Spezia 1929
G-AASJ	*City of Khartoum*	1928	sank, Alexandria 1935
G-AATZ	*City of Swanage*	1928	sold 1937

Short S17 Kent (Scipio)

G-ABFA	*Scipio*	1931	crashed, Mirabella 1936
G-ABFB	*Sylvanus*	1931	burnt out, Brindisi 1935
G-ABFC	*Satyrus*	1931	scrapped 1938

Handley Page HP42

G-AAXC	*Heracles*	1931	wrecked in gale, Whitchurch 1940
G-AAXD	*Horatius*	1931	crashed, Tiverton 1939
G-AAXE	*Hengist*	1931	burnt out, Karachi 1937
G-AAXF	*Helena*	1931	scrapped 1941
G-AAXG	*Hannibal*	1931	crashed in Persian Gulf 1940
G-AAUC	*Horsa*	1931	burnt out, Port Moresby 1940
G-AAUD	*Hanno*	1931	wrecked in gale, Whitchurch 1940
G-AAUE	*Hadrian*	1931	wrecked in gale, Doncaster 1940

Armstrong Whitworth AW15 Atalanta

G-ABPI	*Arethusa*	1932	sold 1933
G-ABTG	*Amalthea*	1931	crashed, Kisumu 1938
G-ABTH	*Andromeda*	1932	withdrawn 1939
G-ABTI	*Atalanta*	1932	to RAF 1941
G-ABTJ	*Artemis*	1932	to RAF 1941
G-ABTK	*Athena*	1933	burnt out, Delhi 1936
G-ABTL	*Astraea*	1933	to RAF 1941
G-ABTM	*Aurora*	1933	to RAF 1941

Short L17 Scylla

| G-ACJJ | *Scylla* | 1934 | to BOAC 1940 |
| G-ACJK | *Syrinx* | 1934 | to BOAC 1940 |

de Havilland DH86

| G-ACPL | *Delphinus* | 1934 | to BOAC 1941 |

G-ACWC	*Delia*	1935	to BOAC 1941
G-ACWD	*Dorado*	1935	to BOAC 1941
G-ADCM	*Draco*	1935	crashed, Austria 1935
G-ADCN	*Daedalus*	1935	burnt out, Bangkok 1938
G-ADFF	*Dione*	1936	to BOAC 1941
G-ADUE	*Dardanus*	1935	to BOAC 1940
G-ADUF	*Dido*	1936	to BOAC 1940
G-ADUG	*Danae*	1936	to BOAC 1940
G-ADUH	*Dryad*	1936	sold 1936
G-ADUI	*Denebola*	1936	to BOAC 1940
G-AEAP	*Demeter*	1936	to BOAC 1940

Short S23 Empire

G-ADHL	*Canopus*	1936	scrapped 1946
G-ADHM	*Caledonia*	1936	scrapped 1947
G-ADUT	*Centaurus*	1936	to RAAF as A18-10, destroyed by enemy 1942
G-ADUU	*Cavalier*	1936	crashed in Atlantic 1939
G-ADUV	*Cambria*	1937	scrapped 1946
G-ADUW	*Castor*	1937	scrapped 1947
G-ADUX	*Cassiopeia*	1937	crashed, Sumatra 1951
G-ADUY	*Capella*	1937	crashed, Batavia 1939
G-ADUZ	*Cygnus*	1937	crashed, Brindisi 1937
G-ADVA	*Capricornus*	1937	crashed, French Alps 1937
G-ADVB	*Corsair*	1937	scrapped 1937
G-ADVC	*Courtier*	1937	crashed, Athens 1937
G-ADVD	*Challenger*	1937	crashed, Mozambique 1939
G-ADVE	*Centurion*	1937	crashed, Bengal 1939
G-AETV	*Coriolanus*	1937	sold 1942
G-AETW	*Calpurnia*	1937	crashed, Habbaniya 1938
G-AETX	*Ceres*	1937	burnt out, Durban 1942
G-AETY	*Clio*	1937	to RAF as AX659 1940
G-AETZ	*Circe*	1937	destroyed by enemy, Java, 1942
G-AEUA	*Calypso*	1937	to RAF as A18-11, 1939
G-AEUB	*Camilla*	1937	sold 1942
G-AEUC	*Corinna*	1937	destroyed by enemy, Australia, 1942
G-AEUD	*Cordelia*	1937	to RAF as AX664, 1940
G-AEUE	*Cameronian*	1937	scrapped 1947
G-AUEF	*Corinthian*	1937	crashed, Darwin 1942

Short S30 Empire

G-AFCT	*Champion*	1938	scrapped 1947
G-AFCU	*Cabot*	1939	to RAF as V3137, 1939
G-AFCV	*Caribou*	1939	to RAF as V3138, 1939
G-AFCW	*Connemara*	1939	burnt out, Hythe 1939
G-AFCX	*Clyde*	1939	wrecked by gale, Lisbon 1941

Short Mayo Composite

S20	G-ADHJ	*Mercury*	1938	scrapped 1941
S21	G-ADHK	*Maia*	1938	destroyed by enemy, Poole 1942

149

Armstrong Whitworth AW27 Ensign

G-ADSR	*Ensign*	1938	scrapped 1945
G-ADSS	*Egeria*	1938	scrapped 1947
G-ADST	*Elsinore*	1938	scrapped 1947
G-ADSU	*Euterpe*	1938	scrapped 1945
G-ADSV	*Explorer*	1938	scrapped 1947
G-ADSW	*Eddystone*	1939	scrapped 1947
G-ADSX	*Ettrick*	1939	damaged by enemy, Le Bourget 1940
G-ADSY	*Empyrean*	1939	scrapped 1947
G-ADZZ	*Elysian*	1939	destroyed by enemy, Merville 1940
G-ADTA	*Euryalus*	1939	written off 1940
G-ADTB	*Echo*	1939	scrapped 1947
G-ADTC	*Endymion*	1939	destroyed by enemy, Whitchurch 1940

de Havilland DH91 Albatross

G-AEVV	*Faraday*	1938	to BOAC 1940
G-AFDI	*Frobisher*	1938	to BOAC 1940
G-AFDJ	*Falcon*	1938	to BOAC 1940
G-AFDK	*Fortuna*	1939	to BOAC 1940
G-AFDL	*Fingal*	1939	to BOAC 1940
G-AFDM	*Fiona*	1939	to BOAC 1940

Short S26 G-Class

G-AFCI	*Golden Hind*	1939	to RAF as X8275, 1940
G-AFCJ	*Golden Fleece*	1939	to RAF as X8274, 1940
G-AFCK	*Golden Horn*	1939	to RAF as X8273, 1940

Other aircraft owned or operated by Imperial Airways

Vickers 66 Vimy Commercial

	G-EASI	*City of London*	from Instone	withdrawn 1925

Vickers 61/74 Vulcan

61	G-EBBL	*City of Antwerp*	from Instone	never operated
74	G-EBFC	[no name]	from Instone	withdrawn 1926
74	G-EBLB	[no name]	from Instone	crashed 1928

Vickers 212 Vellox

G-ABKY	[no name]	1934 (night mail)	crashed 1936

Avro 10

G-AASP	*Achilles*	1930 (charter)	destroyed 1940
G-ABLU	*Apollo*	1931 (charter)	crashed 1938

Westland Wessex

G-AAGW	[no name]	1929 (charter)	sold 1936
G-ABEG	[no name]	1933 (charter)	crashed 1936
G-ACHI	[no name]	1933 (charter)	sold 1936

Boulton Paul P71A

G-ACOX	*Boadicea*	1934	crashed 1936
G-ACOY	*Britomart*	1934	crashed 1935

Avro 652

G-ACRM	*Avalon*	1935	sold 1938
G-ACRN	*Avatar*	1935	sold 1938

British Marine had a third Supermarine Sea Eagle, G-EBFK, written off in a crash on 21st May 1924; in 1925–6 the Supermarine Swan, G-EBJY, was loaned for evaluation on the Guernsey route, but was scrapped in 1927; in 1929–30, an RAF Supermarine Southampton was registered G-AASH and lent to replace the Calcutte *City of Rome*, lost at sea.

Airliners owned and operated by British Airways 1934–40

7 Spartan Cruisers
2 de Havilland DH84 Dragons
11 de Havilland DH86 Dianas
11 de Havilland DH89 Rapides
2 Fokker FVIII
6 Fokker FXII
3 Junkers JU52
7 Lockheed 10 Electra
9 Lockheed 14 Super Electra

Delivered to Imperial Airways in 1934 this Short Scylla was damaged beyond repair when it was blown over in a gale at Drem aerodrome in Scotland. *Captain John Liver.*

Details of airliners operated by Imperial Airways, 1924–40

Biplanes

	max. wt (tons)	passengers	crew & stewards	engines	cruising speed (mph)
Armstrong Whitworth					
Argosy	8	20	2+1	3 AS Jaguar	95
de Havilland					
DH43	3.21	8	2	1 Napier Lion	95
DH50	1.87	4	1	1 AS Puma	95
DH66 Hercules	7	8	2	3 Br Jupiter	95
DH86	4.5	10/14	2	4 DH Gipsy	145
Handley Page					
W8B	5.5	14	2	2 RR Eagle	85
W8F	5.5	12	2	1 RR Eagle	85
				2 AS Puma	
W9	6.5	14	2	3 AS Jaguar	95
W10	6	14	2	2 Napier Lion	90
HP42	13	18/38	3+2	4 Br Jupiter	90
Short					
L17 Scylla	14.5	39	3+2	4 Br Jupiter	95

Monoplanes

Armstrong Whitworth					
AW15 Atalanta	9.5	10	2	4 AS Serval	120
AW27 Ensign	23	27/40	3+2	4 Wright Cyclone	145
de Havilland					
DH91 Albatros	13	22	2+1	4 DH Gipsy	195

Flying-boats

Short					
S8 Calcutta	9	15	3+1	3 Br Jupiter	85
S17 Kent (Scipio)	14	15	3+1	4 Br Jupiter	95
Mayo Composite		(solo/compo)			
S20 floatplane					
(Mercury)	6.9/12	(mail only)	2	4 N-H Rapier	170
S21 flying-boat					
(Maia)	17/7.3	(none)	2	4 Br Pegasus	165

S23 Empire	20	17/24	3+2	4	Br Pegasus	165
S26 G-Class	32.8	40	4+2	4	Br Hercules	180
S30 Empire	23.5	20	3+2	4	Br Perseus	165
Supermarine						
Sea Eagle	2.6	6	1	1	RR Eagle	84

abbreviations
AS Armstrong Siddeley
Br Bristol
DH de Havilland
NH Napier-Halford
RR Rolls-Royce

Limousines arrive at Croydon Aerodrome. *Croydon Airport Society.*

Management of Imperial Airways, 1924–40

Chairmen

Sir Eric Geddes	1924–37
Sir George Beharrel	1937–38
Sir John Reith	1938–40

Government Directors

Sir Herbert Hambling	1924–31
Major J. W. Hills	1924–25
Air Vice-Marshal Sir V. Vyvyan	1925–33
Sir Walter Nicholson	1931–37
Air Chief Marshal Sir J. Salmond	1933–38
Sir Francis Joseph	1937–40
Sir James Price	1938–40

Company Directors

Lord Chetwynd	1928–29
Sir Samuel Instone	1924–37
Sir Hardman Lever	1934–40
Sir George Beharrel	1924–40
H. Scott-Paine	1924–40
I. C. Geddes	1929–40
Hon. E. Harmsworth	1934–40
*G. E. Woods Humphery	1925–38
*Col. F. Searle	1924–25
Lt.-Col. J. Barrett-Lennard	1924–35

(Managing Directors)

Directors of British Airways 1936–40

Hon. Clive Pearson	1936–40	*(Chairman)*
W. D. L. Roberts	1936–40	*(Vice-Chairman)*
Capt. H. Balfour	1936–38	
J. R. Bryans	1936–38	
G. d'Erlanger	1936–40	
E. L. Granville	1936–40	
F. W. Jones	1936–37	
W. C. Tomlinson	1936–40	
J. R. McCrindle	1936–40	*(Managing Director)*
Viscount Monsell	1937–40	*(Government Director)*
E. H. Murrant	1937–40	
Sir R. Chadwick	1938–40	

Secretaries of State for Air 1919–40

Winston Churchill	1919–21
Capt. F. Guest	1921–22
Sir Samuel Hoare	1922–24
Lord Thomson	1924
Sir Samuel Hoare	1924–29
Lord Thomson	1929–30
Lord Amulree	1930–31
Lord Londonderry	1931–35
Viscount Swinton	1935–38
Sir Kingsley Wood	1938–40

The Junkers 52 operated by British Airways in 1937 was built in Germany.

Captain Hinchliffe and the Hon. Elsie Mackay

Lord Inchcape's daughter had engaged the pilot, Captain Hinchliffe to buy a suitable aircraft and agreed to pay him eighty pounds a month whilst preparations were made for the Atlantic crossing. She wanted for herself only the distinction of being Hinchliffe's sponsor and passenger. The prize money of £10,000 would go to him.

Recognising the dangers involved Elsie Mackay undertook to insure the pilot's life for that same sum. Before their take-off she wrote cheques to cover his salary and the insurance premium. Her father was abroad at the time and she was most anxious that her intentions should not become known to him. When the news broke that the aircraft and its occupants must be presumed lost Lord Inchcape, acting as his daughter's trustee, successfully froze her bank account, thereby denying to Hinchliffe's wife the funds due to her.

Sir Sefton Brancker took up Mrs. Hinchliffe's case persuading both Lord Beaverbrook and the proprietor of the Northcliffe press, through their newspapers, the *Daily Express* and *Daily Mail*, to publicise her plight. Lord Inchcape did not respond. The Royal Air Force denied any liability, pointing out that the pilot was no longer a serving officer, nor a member of the RAF reserve.

As the pressure on Lord Inchcape increased he set up a Mackay fund which would gather interest for fifty years and then be offered as a donation to the British government to reduce the national debt. Winston Churchill was Chancellor of the Exchequer and, while welcoming the generosity of the gift, told Mrs. Hinchliffe's sympathisers in the House of Commons that he had no authority to divert any grant to her.

Meanwhile a medium, Mrs. Earl, claimed to be receiving messages from the dead pilot, notifying her of his wife's telephone number and address and begging her to inform the widow that before very long she would receive compensation. Mrs. Earl was most reluctant to increase the anguish of Mrs. Hinchliffe by such an intrusion. She decided to inform Sir Arthur Conan Doyle who was closely involved in spiritualism. Together they approached Mrs. Hinchliffe who attended a seance and was satisfied that the messages did indeed come from her husband.

Finally Lord Inchcape placed £10,000 at the disposal of Winston Churchill to be devoted to 'sufferers from the disaster', a clear reference to Mrs. Hinchliffe and her three children.

Bibliography

W. Armstrong, *Pioneer pilot*, Blandford Press 1952
D. Beaty, *The Water Jump*, Secker & Warburg 1976
F. Brackley, *Brackles*, F. Brackley 1952
Cluett, Nash & Learmonth, *Croydon Airport*, Sutton Libraries 1977
R. E. G. Davies, *Rebels and reformers of the airways*, Smithsonian Institute 1987
E. K. Gann, *Ernest K. Gann's Flying Circus*, Hodder & Stoughton 1974
R. Higham, *Britain's Imperial Air Routes*, Foulis 1960
R. Jackson, *The sky their frontier*, Airlife 1983
J. Lock & J. Creasey, *The log of a merchant airman*, Stanley Paul 1943
C. Martin-Sharp, *D. H.: a history of de Havilland*, Airlife 1960
V. E. Mearles, *Highways of the air*, M. Mason 1948
D. Middleton, *British aviation: a design history*, Ian Allan 1986
G. Olley, *A million miles in the air*, Hodder & Stoughton 1934
I. & R. Ormes, *The skymasters*, Kimber 1976
H. Penrose, *British aviation: the adventuring years*, Putnam 1973
H. Penrose, *British aviation: widening horizons*, HMSO 1979
H. Penrose, *Wings across the world*, Cassell 1980
J. Pudney, *The seven skies*, Putnam 1959
J. C. W. Reith, *Into the wind*, Hodder & Stoughton 1949
J. Stroud, *Railway Air Services*, Ian Allan 1987
Templewood [Sir S. Hoare], *Empire of the air*, Collins 1957
D. W. Wragg, *Boats of the air*, R. Hale 1984
The Times
The Economist
The Aeroplane
Imperial Airways Gazette
Hansard (House of Commons debates)
British Airways archives

General Index

A

Aboukir, 44
Abu Dhabi, 45
Admiralty, 2
Aerodrome Hotel, Croydon, 30, 71
Aero-Lloyd, 15, 26
"Aeroplane" Magazine, 10, 30, 44, 47, 49, 56, 63, 120, 123, 140, 141
Aeropostale, 89–91, 103
Agadir, 93
Aircraft Manufacturing Co, 1
Aircraft Transport and Travel, 1, 3, 10, 11, 13, 36, 77
Air France, 91
Air Ministry, 2, 3, 11, 13, 16, 17, 22, 30, 31, 38–41, 43, 52–54, 60–62, 68, 72, 75, 78, 84, 87, 90, 92, 95, 98, 104–108, 116, 123–126, 128, 130–134, 137–139, 141, 145
Air Orient, 49, 79, 88
Akyab, 40
Alcock, J. Capt, 1, 102
Alderson, M. Capt, 112, 113
Aleppo, 40
Alexander Bay, 111
Alexandria, 43, 46, 49, 54, 56, 85–87, 95–97, 100, 112
Allahabad, 40, 79
American Export Airlines, 111
Amery, L. M.P., 44
Amherst, Lord, 118
Amman, 39
Amsterdam, 14, 15, 25, 35, 36, 73
Amulree, Lord, 104
Anderson, Sir John, 133
Anglo Persian Oil Co, 40
Ankara, 40
Aotearoa, 101
Armstrong, Capt. W., 36, 49, 66, 88
Armstrong-Whitworth, 75, 77, 78
Asquith, H. M.P., 1
Astraea, 85
Atalanta, 82
Athens, 46, 49, 50, 87, 96, 98
Atlee, Clement, M.P, 128

Auckland, 101
Australia, 101
Australian Empire Airways, 82, 83
Australian National Airways, 81
Awarua, 101
Azores, 90, 102, 103, 107–109, 111, 114, 115, 142

B

Baghdad, 33, 39–44, 46, 49
Bahrein, 51, 88
Bailey, Capt. J. F., 24, 95
Baldwin, S. M.P., 27, 44, 73
Balfour, Capt. H. (later Lord Balfour of Inchrye), 51, 66, 73, 81, 134, 144, 154
Balfour, Lord (of Burleigh), 85
Ball, Dr J., 39
Baltimore, 112
Bangkok, 81, 88
Banks, Sir D., 76
Barcelona, 103
Barnard, Capt. F., 6, 22, 24, 27, 34, 35
Barnett Lennard, Col., 18, 154
Basle, 26, 45, 46, 54, 70
Basra, 45, 48, 49, 51
Batavia (now Jakarta), 79
Bathurst (now Banjul), 91–94
Bathurst Island, 81, 82
Beagle, HMS, 86
Beatty, Sir E., 146
Beaverbrook, Lord, 146, 156
Beharrel, Sir G., 18, 118, 123, 129, 132, 134, 135, 140, 154
Beirut, 49
Beit, Sir Alfred, 54
Belfast, 43, 70
Benghazi, 44
Bennett, Capt. D. (later Air Vice Marshal), 111, 112, 115
Berlin, 14, 15, 17, 25, 36, 40, 72, 88, 90, 91, 102, 112
Bermuda, 90, 99, 102, 104, 106–109, 112, 113
Biggin Hill, 35
Blackburn Aircraft Co, 37, 52–54

Bleriot, L., 1
Bombay, 41
Bonar Law, A., M.P., 15, 16
Bordeaux, 92, 95
Botwood, 108–111, 116
Boulama, 94
Boulogne, 6
Bowen, J. W., 128
Brackley, Major H., 4, 7, 10, 23, 28, 32, 33, 36, 46, 48–50, 58, 61, 71, 77, 81–83, 86, 95, 100, 109, 119, 120, 141, 144, 146
Brackley, Mrs. F., 32
Brancker, Sir Sefton, 1, 3, 4, 9, 10, 13, 22, 37, 39–42, 47, 53, 54, 62, 79, 117, 156
Bremen, 103
Brindisi, 46, 49, 50, 59, 60, 71, 85, 96, 98
Brisbane, 84
British Airline Pilots Assoc, 118–123, 125, 128, 133, 139
British Airways Ltd., 73, 75, 76, 78, 92, 118, 123–126, 128, 134, 140, 141, 144
British Broadcasting Corp., 117, 129, 131, 135, 136, 140
British Continental Airways, 70, 73
British Marine Air Navigation Co., 14, 18, 25, 95
British Overseas Airways Corp., 36, 61, 70, 117, 141, 145, 146
British Purchasing Corp, 146
British South American Airways, 146
Brown, Harold, M.P., 143, 146
Brown, Sir William, 128
Bruges, 68
Brussels, 5, 6, 13, 14, 16, 25, 61, 70, 126
Bucharest, 40
Budapest, 46, 70, 72, 120–122, 125, 126
Buenos Aires, 91
Bullock, Sir C., 75
Burchall, Col. H., 47, 56, 61, 81, 120, 136, 146
Burney Airship Scheme, 52
Burslem Lane, Capt, 119, 121, 122, 139
Bushire, 44

C
"C" Class Flying Boats Names:
　Cabot, 115, 143
　Caledonia, 95–97, 108, 109
　Calpurnia, 98
　Calypso, 97, 99

Cambria, 97, 108
Camilla, 101
Canopus, 60, 95, 96
Capella, 101
Capricornus, 98
Caribou, 115, 143
Castor, 96, 98
Cathay, 115
Cavalier, 96, 99, 100, 109, 112, 113
Centaurus, 96, 101
Centurion, 97, 99
Ceres, 97, 98
Challenger, 99
Champion, 100, 113
Clyde, 100, 115
Connemara, 99, 115
Coorong, 101
Corsair, 98–100
Courtier, 98
Cygnus, 98
Cadman Committee, 128, 130–134, 139
Cairo, 33, 39, 40, 43, 46–48, 50–54, 56, 57, 75, 78, 97, 125
Calcutta, 40, 41, 49, 61, 71, 79, 84, 99
Campbell-Black, J., 54, 83
Campbell-Orde, A., 76, 77, 141
Canadian Pacific Railways, 146
Candia, 50
Canton, 88
Cape Town, 22, 47, 52–54, 56–59
Cardiff, 66
Cardington, 10, 103
Casablanca, 93
Chamberlain, N., M.P., 128, 133–135
Chamberlin, C., 102
Cherbourg, 99
Chesham, Lord, 118
Chetwynd, Lord, 46, 81, 154
Churchill, W. S., M.P., 3, 9, 11, 39, 155, 156
City of Cairo, 42, 44
City of Delhi, 44
City of Khartoum, 85, 120, 126
City of Rome, 46
City of Stonehaven, 86
Clacton-on-Sea, 69
Clipper III, 109, 110
Cobham, Sir Alan, 40, 53, 54, 81, 105, 115, 123
Cologne, 14, 15, 17, 25, 40, 49, 61, 68, 72, 73, 76, 81, 121, 126

Conakry, 94
Constantinople (now Istanbul), 40
Cook, Thomas, 61
Copenhagen, 10, 73
Corfu, 46
Corsairville, 100
Coster, A. J., 116
Cowdray, Lord, 66, 120
Cowes, 16, 70
Cricklewood Aerodrome, 56
Crilly Airways, 73, 74
Cripps, Capt., 75
Croydon Airport, 8, 10–13, 15, 17, 21, 27–30, 35, 36, 40, 42, 56, 60, 61, 65, 66, 68, 70, 75, 76, 80, 81, 83, 84, 96, 117, 120, 123, 124, 126, 135, 143
Cumberpatch Trophy, 120
Cunard Shipping Co., 9
Cyprus, 43, 50

D
Daily Chronicle Newspaper, 11
Daily Mail Newspaper, 1, 65, 156
Daimler Airway, 11–15, 17, 18, 21, 25, 35
Dakar, 9, 89, 93
Daladier, E., 134
Damascus, 50
Dangu River, 99, 100
Darwin, 80–82, 84, 87, 100, 101
De Havilland Aircraft Co., 40, 67, 70, 108, 139
De Havilland, Geoffrey, 1, 77
Delhi, 39, 44, 46, 49, 79, 83
D'Erlanger, G. (later Sir Gerard), 70, 73, 120, 124, 128, 154
De Valera, 109
Dijon, 44
Dingari Lake, 98
Dismore, Capt., 24, 35, 63, 65
Dismore, S. A., 58, 136, 146
Douglas Aircraft Corp., 112
Dubai, 45
Dundee, 111
Dunlop Rubber Co., 13, 19, 118, 126
Durban, 59, 60, 97

E
Eckener, H., 102
Economist Magazine, 123, 140
Edenbridge, 12

Edinburgh, 38
Egeria, 78
Eisleben, 102
El Fasher, 93, 94
El Jid, 39
Elsinore, 78
Empire Air Mail Scheme, 13, 58, 60, 95–97, 101, 107, 118, 132
Esfahan, 40
Esso Baytown, 113
Euphrates River, 51
Europa, 103
Euterpe, 78
Exeter Airport, 143

F
Fisher, Sir Warren, 72, 92, 134, 136
Flight Refuelling, 105, 115
Flying Scotsman, 38
Fokker, 10
Forbes, Lord, 139
Fort Lamy, 93
Foynes, 108–110, 116
Franco, General, 75, 92, 93
Frankfurt, 121
Freetown, 94
Friedrichshafen, 102, 103
Fysh, Sir Hudson, 93, 100, 136

G
Galway, 1
Gander, 115
Gatwick Airport, 75, 76, 78, 124
Gaya, 79
Gaza, 48
Geddes, Sir Eric, 13, 18, 19, 21–23, 29, 30, 32, 37, 46, 48, 57, 58, 62, 65, 67, 68, 70, 72, 76, 83, 84, 117, 118, 131, 132, 136, 138, 139, 154
General Post Office, 9, 58, 70, 89, 92–94
Genoa, 42, 46, 48–50
George V, King, 4
Gibraltar, 73
Gladstone, Tony, 53, 56
Glasgow, 70
Golden Fleece, 116
Golden Hind, 116, 142, 143
Golden Horn, 116
Granville, K. (later Sir Keith), 61, 117
Gravesend, 124

Gray, Capt., 110, 114
Great Western Railway, 66, 67
Grey, C. G., 30, 34, 47, 78, 120, 131, 141
Greyhound Hotel, 118
Gsell, Father, 82
Guam, 107
Guernsey, 26, 35, 43
Guest, Capt. F., M.P., 11, 13, 43, 155
Guild of Air Pilots and Navigators, 117, 120, 122, 125
Gwadar, 51

H
Habbaniyah, 98
Hague, 10
Haifa, 49
Halle, 70
Hambling, Sir H., 16, 18, 72, 121, 136, 154
Hamburg, 10
Handley Page Transport, 4–7, 10, 11, 14, 17, 18, 21, 22, 25, 26, 30, 38, 89, 119
Hannibal, 57, 59, 63
Hanover, 25, 73
Hatties Camp, 115, 116
Heliopolis, 43, 48
Hendon, 3
Hengist, 60, 77
Heracles, 62
Heston Aerodrome, 78, 92, 124, 126, 144
Highland Airways, 73
Hinchliffe, Capt. W. R., 17, 24, 28, 34, 35, 102, 156
Hinkler, Bert, 81
Hitler, Adolf, 73, 112, 134, 142
Hoare, Sir S., M.P., 16, 18, 27, 29, 36, 44, 53, 54, 79, 155
Holt Thomas, G., 1, 9
Homs, 44
Hope, Capt. W., 88, 109
Hore-Belisha, L., M.P., 134
Horsa, 88
Horsey, Capt. A., 24, 35, 72
Hounslow Airport, 3, 4, 6, 8
Hughes, H., 128
Hull, 37
Hythe, Hants, 96, 98, 100

I
Ifould, Lloyd, 32, 33, 69
Ile de France, 103

Irchcape, Lord, 35, 156
Indian Airways, 79
Indian Trans-Continental Airways, 81
Instone Air Line, 10, 11, 13, 14, 17, 18, 22, 25, 29, 31, 36
Instone, Sir Samuel, 6, 20–22, 154
Instone, Theodore, 4
Into the Wind, 129

J
Jaffa, 50
Jask, 44, 47
Jodhpur, 44
Johannesburg, 56
Johnson, Amy, 81
Johnston, T., M.P., 16, 17
Jones, Capt. O. P., 17, 24, 34, 72
Juba, Sudan, 59, 99, 100
Junkers Co., 41, 44, 52, 59

K
Kaduna, 93, 94
Khan of Kulat, 44
Kano, 93, 94
Karachi, 41, 44, 49, 51, 77–81, 83, 97
Karnak, 59
Kerman, 40
Khartoum, 50, 53, 55, 56, 93, 94, 97
Kidston, G., 56
King's Cup Air Race, 35
Kingsford Smith, Sir C., 81, 83–85
Kisumu, 52–54, 88
K.L.M. (Royal Dutch Airlines), 6, 15, 36, 51, 65, 74, 80, 81, 83, 95, 126
Koepang, 80, 81
Konigsberg, 7

L
Lagos, 93, 94
Lahore, 40
Lakehurst, USA, 102, 103
Langstone Harbour, 132
Las Palmas, 89, 91
Le Bourget, 3, 4, 7, 31, 32, 64, 144
Leipzig, 70
Leopoldville, 30
Le Touquet, 75
Le Zoute, 70
Lille, 70
Lindbergh, Col., 35, 102, 104, 106, 107

Lingeh, 94
Lisbon, 75, 89, 92, 93, 109, 111, 114, 115
Liverpool, 43, 70, 89
Lloyd-George, M.P., 1, 15
Lloyd Triestino Steamship Co., 50
Londonderry, Lord, 72, 106, 155
Lorenz Receiver, 75, 76
Lowenstein Wertheim, Princess, 35
Lufthansa, 34, 36, 88, 91, 112
Luxor, 59
Lympne, 12, 17, 37
Lynch Blosse, Capt. P., 70
Lyons, 98

M
Macao, 88
Macdonald, R., M.P., 76
Mackay, Hon. Elsie., 35, 156
MacKinnon, Sir P., 90
Madeira, 103, 109
Maidugari, 93
Maitland, Brigadier E., 1, 10
Malakal, 59
Malmo, 70, 73
Manchester, 14, 34
Marconi, 11, 75
Marden, 12
Marquis, Sir F. (Later Lord Woolton), 128
Marseille, 42, 44, 60, 91, 95, 114
May, P., 61
Mayo, Major R., 60, 68, 84, 105, 111, 138
McCrindle, Major J. R., 70, 73, 144, 154
McIntosh, Capt. R., 7, 10, 11, 12, 24, 35
Melbourne, 81–84
Mersah Matruh, 46
Midway Island, 107
Milan, 49
Mildenhall to Melbourne Air Race, 70, 83
Minchin, Lt. Col. F., 28, 35, 41, 102
Minna, 94
Mirabella, 46, 86, 87
Mollard, Capt. R., 80, 81, 87, 98
Moncton, 116
Montague, F., M.P., (Later Lord Amwell), 127
Montreal, 107, 109–112
Moore Brabazon, Lt. Col. J., (Later Lord Brabazon), 126
Morning Post Newspaper, 27
Moscow, 7

Muirhead, Lt. Col. A., M.P., 121, 128
Muscat, 45
Mussolini, B., 46, 49, 93
Mwanza, 50, 54, 56

N
Nairobi, 52, 54, 56, 59
Nanking, 88
Naples, 44, 49, 98
Natal, Brazil, 91
Neufchatel, 62
New York, 99, 100, 102, 104, 108–112, 141
Norman, Sir Montagu, 136
North Atlantic Ferry Org., 146
North British Locomotive Co., 129
Northcliffe, Lord, 1, 4, 22, 102
North-East Airways, 124
Northern & Scottish Airways, 73
North European Grand Trunk Airway, 15

O
Observer Newspaper, 136
Olley Air Services, 5, 69, 118
Olley, Capt. Gordon, 5, 24, 38, 66, 69
Oshogbo, 94
Oslo, 10, 144
Ostend, 25, 70

P
Pan American Airways, 100, 104, 106–111, 113, 114, 115, 141, 142
Paris, 3–6, 10, 11, 13–16, 26, 35, 36, 40, 42, 44, 46, 49, 60, 61, 69, 70, 72, 77, 78, 102, 121, 123, 126, 131, 139
Pasni, 49
Peace Conference 1919, 9
Pearson, Hon. Clive, 66, 120, 140, 142, 146, 154
Peking (now Beijing), 7
Penang, 81, 88
Peninsula & Oriental Line, 42
Penshurst, 12
Perkins, R., M.P. (Later Sir Robert), 120, 121–126, 129, 133
Pernambuco, 90
Perth, Australia, 81
Perth, Scotland, 144
Pick, F., 2
Pietersburg, 59
Pilkington, Col. W., 3

Pioneer Pilot, 66
Pisa, 44
Plymouth, 66
Poole, 143
Porte, Capt J., R.N., 2, 105
Port Etienne, 93
Port Washington, 111
Powell, Capt. G. J., 24, 48, 57, 112
Prague, 9, 14, 29, 70, 121
Price, Sir J., 136, 139, 154
Purley, 27

Q
Qantas Empire Airways, (Q.E.A.), 83, 84, 95, 100, 101
Queensland & New Territories Air Services, 83
Quetta, 40

R
Rae Committee, 131
Railway Air Services, 67, 69, 120
Rambang, 80
Rangoon, 40, 41, 49, 79, 81, 83
Reith, Sir J. (later Lord Reith), 129, 130, 134–136, 139, 141, 142, 144, 154
Rhodes, Sir C. 54
Rhodes, 50
Rhodesia & Nyasaland Airways (R.A.N.A.), 59
Rineanna, 108
Rio de Janeiro, 89–91, 146
Robertson, Sir M., 83
Robinson, Capt., 24, 72
Rogers, Capt. Kelly, 100, 115, 116
Rogers, Capt. W., 10, 24, 65, 66, 119, 121, 122, 139
Rome, 49, 95, 96
Romford, 69, 70
Ropner, Col., M.P., 127
Rotterdam, 17
Royal Aeronautical Soc., 30
Royal Air Force, 1, 7, 20, 21, 39, 41, 43, 46, 51–53, 57, 69, 71, 73, 78, 79, 88, 100, 105, 114–116, 119, 124, 137, 143, 144, 146
Royal Flying Corps, 4, 36, 66
Royal Naval Air Service, 77
Runciman, Hon. L. (later Lord Runciman), 143, 146
Rutbah Wells, 41, 43

S
Sabena, 30, 71
Saigon, 49, 88
Salisbury, Rhodesia (now Harare), 56, 59
Salonika, 46
Salmond, Maj. Gen. G., 39
Salmond, Sir J., 98, 133, 154
Sassoon, Sir P., M.P., 60, 107
Satyrus, 97
Saunders Roe, 51, 66
Savoy Hotel, 25
Scipio, 50, 87
Scott, C. W. A., 83
Scott-Paine, H., 18, 144, 154
Scylla, 68, 69, 75, 77
Sea of Galilee, 50
Searle, Col. F., 11, 18, 19, 22, 23, 25, 29, 154
Seville, 91
Shanghai, 88
Shannon, 115
Sharjah, 45, 51
Shaw, J., 3, 7
Sheikh Issa, 51
Shell Petroleum Co., 59
Shelmerdine, Lt. Col. F., 54, 60, 77, 106, 107
Sholto-Douglas, Lt. Col., (later Lord Douglas), 5
Short Bros., 60, 69, 95, 111, 139, 144
Silver Wing, 35, 63
Singapore, 79–84, 97, 98, 100
Skoplje, 46
Smith, K. & R., 39
Smuts, Gen. Jan., 57
Soc. of British Aircraft Constructors, 40
Sollum, 44
South African Airways, 97
Southampton, 14, 43, 92, 96, 98, 111, 114, 115, 143
Southern Railway, 66, 70, 98
Spartan Airlines, 70, 73
Spezia, 46
Stainton, J. R. (later Sir Ross), 117
Standard Beam Approach (SBA), 76, 121
Stockholm, 10, 144
Store, Capt. G., 115
St. Paul Island, 89
Strasbourg, 40
Stuttgart, 40, 91
St. Vincent, 89

Suda Bay, 46
Surrey Flying Services, 118
Swinton, Viscount, 72, 76, 78, 121, 128, 134, 155
Swissair, 70, 126
Sydney, Australia, 80, 101
Sykes, Maj. Gen., 2, 7, 102
Sylvanus, 50
Syrinx, 68, 75, 77
Szarasy Corp., 16, 17

T
Takoradi, 94
Tangye, N., 122
Tasman Empire Airways, 95, 101
Teheran, 7, 41, 44, 45
Thomson, Lord, 21, 22, 27, 62, 155
Tigris River, 51
Tillett, Ben, 21
Times Newspaper, 121, 122, 140
Tobruk, 46
Tonbridge, 12, 63
Toulouse, 89
Trades Union Congress, 21, 31, 32, 118
Transport & General Workers Union, 127
Treaty of Versailles, 7
Trenchard, Lord, 2, 39, 104, 138
Trippe, Juan, 104, 106, 107, 111, 114, 141, 142

U
Union Airways, 59

V
Valley of the Kings, 59
Vancouver, 109

Verney, Air Comm., 106, 107
Vickers Aircraft Co., 90, 103, 104, 107, 138
Victoria Point, 79
Vienna, 121

W
Waddon, 8 , 65
Wake Island, 107
Wakefield, Sir C., 54
Wallington, 8
Walters, Capt., 49
Warsaw, 40
Wells, H. G., 12
Weir, Lord, 9
Westfalen, 91
Whitchurch, 143
Whitehall Trust Ltd., 66, 73, 102, 124, 128
Whitten Brown, 1, 102
Wilcockson, Capt., 24, 49, 109, 110, 136
Williams, Bob, 21
Wilson, Capt., 120
Wilson, Sir Horace, 134
Woods Humphery, G., 4, 11, 15, 19, 21–24, 29, 30, 32, 44–48, 57, 58, 62, 71, 76, 81, 83, 93, 104, 106, 107, 111, 119–122, 128–132, 134–142, 144, 146, 154
Wolley Dod, Capt., 24, 33, 41, 47, 54, 76
Wood, Sir Kingsley, 78, 134, 140, 155

Y
Yankee Clipper, 114, 141
Youell, Capt., 24

Z
Zurich, 26, 61

Index of Aircraft

Armstrong-Whitworth
    Argosy, 30, 31, 34, 35, 36, 42, 45, 46, 56, 75, 80
    Atalanta, 56, 57, 58, 65, 77, 81–85, 87
    Ensign, 75, 77, 78
    Whitley, 75
A. V. Roe
    Avro 10, 68

Boeing
    Boeing 314, 63, 110, 111, 113, 114, 116, 142
De Havilland
    D.H.80A Puss Moth, 56, 69
    D.H.83 Fox Moth, 69
    D.H.4, 3
    D.H.9, 3, 52

D.H.18, 11, 12, 13
D.H.34, 13, 15, 17, 25, 27, 36
D.H.50, 40, 53
D.H.66 Hercules, 33, 34, 42–44, 47, 49, 56, 70, 79–81, 87
D.H.84 Dragon, 67, 69, 83
D.H.86 Express, 60, 75, 76, 83, 84, 88, 94, 122
D.H.89 Rapide, 67, 143
D.H.91 Albatross, 77, 114, 142
D.H.95 Flamingo, 78
Dornier
  D.O.18E, 91
  WAL, 91
Douglas
  D.C.2, 70, 83, 95, 112
  D.C.3, 63
  D.C.4, 112
Fairey
  Fairey III, 52–54, 89, 95, 112
Focke Wulf
  Kondor, 112
Fokker
  F.VIII, 73
  F.XII, 75, 76
Handley Page
  0/400, 4–6
  W.4, 6
  W.8, 7, 62, 65
  W.8.B, 25, 35, 36
  W.8.F, 26
  W.9.A, 30, 35
  W.10, 29–31, 35, 46
  H.P.42, 38, 51, 56, 57, 60, 62, 63, 65, 69, 78, 87, 143
  Harrow, 138

Heinkel
  H.E.70, 91
Junkers
  G.24, 30
  J.U.52, 73, 144
Lockheed
  Altair, 85
  10 Electra, 73
  14 Super Electra, 78, 92, 104
  Hudson, 146
  Vega, 57
Martin
  130, 106, 107
Short Bros.
  Calcutta, 43, 45, 46, 48–51, 56, 85–87, 97
  "C" Class, 60, 95–97, 98, 99, 100, 101, 108, 109, 112, 113, 115, 143
  "G" Class, 111, 114, 116, 141, 142, 144
  Kent, 49, 50, 56, 69, 85, 87, 97
  Maia, 105, 111
  Mercury, 84, 105, 111, 112
  Rangoon, 51
  Scion, 94
  Scylla, 68, 69
  Singapore, 43, 54
  Syrinx, 68
Sikorsky
  S.42, 84, 106, 109–111
Spartan Cruiser, 70
Supermarine
  Sea Eagle, 14, 26, 35
Vickers
  Vimy, 1, 25, 102
  Vulcan, 25, 36
Westland
  Wessex, 66

Index of Airships

Graf Zeppelin, 90, 103, 104
L.Z.126, 102
R.34, 1, 102

R.38, 10
R.100, 10, 90, 103, 104
R.101, 10, 52, 62, 79, 90, 103, 104